PRINCIPLES OF SOIL AND WATER ANALYSIS

NEW INDIA PUBLISHING AGENCY
New Delhi-110 034

About the Author

Dr. Narayan Panda obtained his B.Sc.(Ag) degree with Hons in the year 2001 from Odisha University of Agriculture & Technology (OUAT), Bhubaneswar, Odisha. He received his M.Sc.(Ag) in Soil Science and Agricultural Chemistry in the year 2003 and PhD in Soil Science and Agricultural Chemistry in the year 2006 from Banaras Hindu University (BHU), Varanasi, India. He is the recipient of the *Binani Gold Medal* and *BHU Gold medal* award for outstanding academic performance during his post graduate studies. He has been associated with five research projects as Co-PI and edited nine practical manuals on soil science courses for under graduate programme of the University. He has published 26 research papers in National and International Journals and 12 popular articles in different magazines. He has 13 years of experience in teaching, research and extension.

This book is written for both the under graduate and post graduate students of Soil Science discipline. Other persons working in the soil and water testing laboratories may also use this book for laboratory analysis.

PRINCIPLES OF SOIL AND WATER ANALYSIS

Narayan Panda
Assistant Professor
Department of Soil Science and Agricultural Chemistry
College of Agriculture, OUAT
Bhubaneswar-751003

NEW INDIA PUBLISHING AGENCY
New Delhi-110 034

NEW INDIA PUBLISHING AGENCY
101, Vikas Surya Plaza, CU Block
L.S.C. Market, Pitam Pura, New Delhi-110 034
Ph : +91 11 27341616, 27341717, 27341718
E-mail:newindiapublishingagency@gmail.com
Web: www.nipabooks.com

For customer assistance, please contact
Phone: + 91-11-27 34 17 17
Fax: + 91-11- 27 34 16 16
E-Mail: feedbacks@nipabooks.com

ISBN: 978-81-19235-58-2

Composed and Designed by NIPA®.

Preface

Sustainable agricultural management approaches in the soil environment is the key to sustained food security, soil quality and livelihood security. The damage to soil fertility and depletion of plant nutrient elements in soil has emerged as major threats to the sustainability of diverse agricultural production systems. Therefore, soil and water testing is considered an important tool to assess the essential plant nutrient elements need to formulate crop and site specific nutrient recommendations vis-à-vis climate change strategies. Soil test based nutrient use optimizes economic returns and minimizes risks of environmental pollution.

Characterization of soil mineralogical, microbiological or biochemical properties within an experimental area among different treatments or cropping systems it requires the utmost attention to planning details. It is imperative to establish experimental objectives as a prelude to developing the sampling plan. As a component of development of sampling plan, one must begin by evaluating the environment from which samples will come. To accomplish this goal one must determine the physical and chemical features of the experimental area that includes the basic agents of nutrient supply i.e. soil and water.

A need, therefore, was felt to bring out a book on "Principles of Soil and Water Analysis" that can helpful to the Soil Chemists, Agriculture Officers, Specialists and research scholars of soil testing laboratories and other laboratory technicians/assistants. The present book is an attempt in that direction. It contains some principles and reactions of chemical analysis, soil sampling and analytical techniques for the determination of physical and chemical parameters of soil and water samples.

N. Panda

Contents

1

Analytical Chemistry – Basic Concepts

Many chemical reactions in nature, particularly in living systems, occur in aqueous solutions. Reactions are often carried out in the laboratory in aqueous solution because these are generally more rapid in solution than when pure solid reactions are mixed. Analysis of substances may be qualitative or quantitative in nature. A quantitative analysis is one in which the amount or concentration of a particular species in a sample is determined accurately and precisely.

1.1 Volumetric Analysis

Volumetric analysis is a quantitative analysis. It involves the measurement of the volume of a known solution required to bring about the completion of the reaction with a measured volume of the unknown solution.

The reaction used in volumetric analysis must fulfil the following conditions;

1. The chemistry of the reaction should be known and it should be represented by a definite chemical equation.
2. The reaction should be instantaneous and proceed to completion of over a fairly wide range of concentrations.
3. There must be means to know the exact state of its completion.

1.2 Titration

The process of addition of the known solution from the burette to the measured volume of solution of the substance to be estimated until the reaction between the two is just complete is termed as titration.

Titration involves two solutions

1. **Unknown solution:** The solution consisting of the substances to be estimated is termed as unknown solution. The substance is called titrate.
2. **Standard Solution:** The solution in which an accurately known amount of the reagent (titrant) has been dissolved in a known volume of solution is termed standard solution. There are two types of titrants;

a. Primary standards: These can be accurately weighted and their solutions are not to be standardized before the use.

e.g Oxalic Acid ($H_2C_2O_4.2H_2O$), Potassium dichromate ($K_2Cr_2O_7$), Silver nitrate ($AgNO_3$), Copper sulphate($CuSO_4.5H_2O$), Ferrous ammonium sulphate [$FeSO_4(NH_4)_2SO_4.6H_2O$], Sodium thiosulphate ($Na_2S_2O_3.5H_2O$), etc

b. Secondary standards: The solutions of these reagents are to be standardised before use as these can not be weighed accurately.

e.g. Sodium hydroxide (NaOH), Potassium hydroxide (KOH), HCl, H_2SO_4, $KMnO_4$, I_2, etc

1.3 Law of equivalence

It is applied in all volumetric estimations. According to it the chemical substances react in the ratio of their chemical equivalent masses.

$$\frac{\text{Mass of substance A}}{\text{Mass of substance B}} = \frac{\text{Chemical equivalent mass of A}}{\text{Chemical equivalent mass of B}}$$

or, $$\frac{\text{Mass of substance A}}{\text{Chemical equivalent mass of A}} = \frac{\text{Mass of substance B}}{\text{Chemical equivalent mass B}}$$

Or, Gram equivalent of A = Gram equivalent of B

Or, Milli-gram equivalent of A = Milli-gram equivalent of B

The point at which the amount of the two reactants are just equivalent is known as equivalent point or end point. An auxiliary substance which helps in the usual detection of the completion of the titration or equivalence point or end point is termed as an indicator. Indicators are substances which undergo some easily detectable changes at the equivalence point are used as indicators.

1.4 Types of Titration

Titration can be classified as

1. Acid-base titration or acidimetry and alkalimetry
2. Oxidation-reduction titrations or redox titrations
3. Precipitation titrations
4. Complexometric titrations

1. Acid-base titration

When the strength of an acid is determined with the help of a standard solution of base, it is known as acidimetry. Similarly when the strength of a base (alkali) is determined with the help of a standard solution of an acid, it is known as alkalimetry. Both these titrations involve neutralization of an acid with alkali. In these titrations H^+ ions of the acid combine with OH^- ions of the alkali to form unionised molecules of water.

$$HA\ (Acid) + BOH\ (Alkali) \rightarrow BA\ (Salt) + H_2O\ (Water)$$

$$\text{Or, } H^+ + A^- + B^+ + OH^- \rightarrow B^+ + A^- + H_2O$$

$$H^+ + OH^- \rightarrow H_2O$$

The end point in these titrations is determined by the use of organic dyes which are either weak acids or weak bases. These change their colours within a limited range of hydrogen ion concentration i.e., pH of the solution.

2. Oxidation-Reduction Titration

The titrations based on oxidation-reduction reactions are called redox titrations. The chemical reactions proceed with transfer of electrons (simultaneous loss or gain of electrons) among the reacting ions in aqueous solution. Sometimes these titrations are named after the reagent used as;

a). Permanganate titration

These are titrations in which potassium permanganate is used as an oxidising agent in acidic medium. The medium is maintained by the use of dilute sulphuric acid. Potassium permanganate acts as a self indicator. The potential equation, when potassium permanganate acts as an oxidising agent is;

$$2KMnO_4 + 3\ H_2SO_4 \rightarrow K_2SO_4 + 2\ MnSO_4 + 3\ H_2O + 5[O]$$

$$\text{or } MnO^-_4 + 8\ H^+ + 5e \rightarrow Mn^{2+} + 4\ H_2O$$

Before the end point, the solution remains colourless (when $KMnO_4$ solution is taken in burette) but after the equivalence point only one extra drop of $KMnO_4$ solution imparts pink colour, i.e. appearance of pink colour indicates the end point. $KMnO_4$ is used for the estimation of ferrous salt, oxalic acid, oxalates, hydrogen peroxides, etc. The solution of $KMnO_4$ is always first standardised before its use.

b). Dichromate titrations

These are titrations in which potassium dichromate is used as an oxidising agent in acidic medium. The medium is maintained acidic by the use of dilute sulphuric acid. Potassium permanganate acts as a self indicator. The potential equation, when potassium dichromate acts as an oxidising agent is;

$$K_2Cr_2O_7 + 4\ H_2\ SO_4 \rightarrow K_2SO_4 + Cr_2(SO_4)_3 + 4H_2O + 3[O]$$

$$\text{Or, } Cr_2O_7^{2-} + 14\ H^+ + 6e \rightarrow 2Cr^{3+} + 7\ H_2O$$

The solution of potassium dichromate can be directly used for titrations. It is mainly used for the estimation of ferrous salts and iodides. In the titration of $K_2Cr_2O_7$ versus ferrous salt either an external indicator (potassium ferricyanide) or an internal indicator (diphenyl amine) can be used.

*c). **Iodimetric and iodometric titrations***

The reduction of free iodide to iodine ions and oxidation of iodide ions to free iodine occurs in these titrations.

$$I_2 + 2e \rightarrow 2I^- \text{(reduction)}$$

$$2\ I^- \rightarrow I_2 + 2e \text{ (oxidation)}$$

These two are divided into two types;

*(i). **Iodimetric titrations***

These are the titrations in which free iodine is used. As it is difficult to prepare the solution of iodine (volatile and less soluble in water), it is dissolved in potassium iodine solution.

$$KI + I_2 \rightarrow KI_3 \text{ (Potassium tri – iodide)}$$

This solution is first standardised before use. With the standard solution of substances such as sulphite, thiosulphate, arsenite, etc are estimated.

*(ii). **Iodometric titrations***

In iodometric titrations, an oxidising agent is allowed to react in neutral medium or in acidic medium with excess potassium iodide to liberate free iodine.

$$KI + \text{Oxidising agent} \rightarrow I_2$$

Free iodine is titrated against reducing agent usually with sodium thiosulphate. Halogens, oxyhalogens, dichromates, cupric ion, peroxides, etc can be estimated by this method.

$$I_2 + Na_2S_2O_3 \rightarrow 2NaI + Na_2S_4O_6$$

$$2CuSO_4 + 4KI \rightarrow Cu_2I_2 + 2K_2SO_4 + I_2$$

$$K_2\ Cr_2O_7 + 6KI + 7\ H_2SO_4 \rightarrow Cr_2\ (SO_3)_3 + 4K_2SO_4 + 7H_2O + 3I_2$$

In iodimetric and iodometric titrations, starch solution is used as an indicator. Starch solution gives blue or violet colour with free iodine. At the end point the blue colour or violet colour disappears when iodine is completely changed to iodine salts.

3. Precipitation titrations

The titrations which are based on the formation of insoluble precipitates, when solutions of two reacting substances are brought in contact with each other are called precipitation titrations. For examples, when a solution of silver nitrate is added to a solution of sodium chloride or a solution of ammonium thiocyanate, a white precipitate of silver chloride or silver thiocyanate is formed.

$$AgNO_3 + NaCl \rightarrow AgCl + NaNO_3$$

$$AgNO_3 + NH_4CNS \rightarrow AgCNS + NH_4NO_3$$

Such titrations involving silver nitrate are called argentometric titrations.

4. Complexometric titrations

A titration in which an un-dissociated complex is formed at the equivalence point is called complexometric titraion. These titrations are superior to precipitation titrations as there is no error due to co-precipitation.

$$Hg^{2+} + 2\ SCN^- \rightarrow Hg(SCN)_2$$

$$Ag^+ + 2CN^- \rightarrow AgCN$$

EDTA (ethylenediamine tetra-acetic acid) is a useful reagent which forms complexes with metals in the form of disodium salt. It is used to estimate and ions in presence of erichrome black-T (EBT) as an indicator.

1.5 Types of Indicators

Indicators are the complex organic compounds of high molecular weight which exhibit different colours under different conditions. There are three types of indicators

a. Acid base indicators

b. Redox indictors

c. Metal ion indicators

a) **Acid base indictors** exhibit different colours in acid and alkaline medium. In water and other solvents they behave as weak acids or bases and participate in equilibrium reactions involving H^+. The change in colour is due to association or dissociation of protons accompanied by complex internal structural rearrangements.

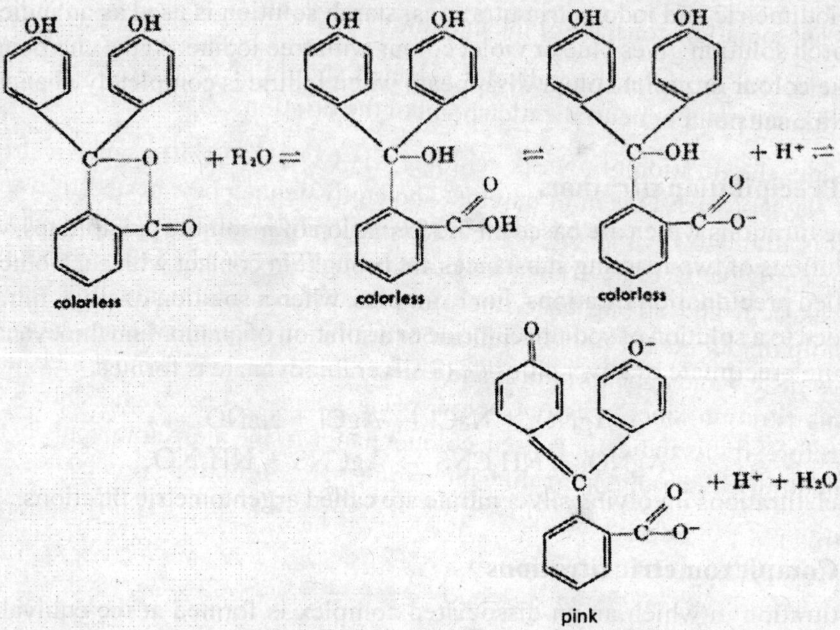

An example is the phenolphthalein indicator. It is colourless below pH 8.5, whereas it exhibits pink colour above pH 8.5. The reaction occurs as above.

It can be seen that phenolphthalein undergoes different forms with structural changes to exhibit pink colour from the colourless form. Equilibrium exists among different forms of phenolphthalein. When the equilibrium concentration of the colourless form of the indicator is much higher than the concentration of the form exhibiting pink colour, the indicator appears colourless and the pink colour appears when the reverse occurs. On adding a base, the H^+ is neutralized. So the equilibrium proceeds in the forward direction. The concentration of the colourless form then decreases, whereas the concentration of the form exhibiting pink colour increases. On adding an acid, concentration of H^+ increases. Consequently the rate of the reverse reaction increases and the equilibrium shifts from right to left, i.e. the concentration of the pink colour species decreases and the concentration of colourless species increases.

For simplicity, without discussing the structural rearrangements, the association and dissociation reaction of the acid-base indicators can be represented as follows;

$$\underset{\text{Acid colour}}{HIn} \rightleftharpoons \underset{\text{Base colour}}{H^+In^-}$$

$$K_{in} = \frac{[\mathrm{H}^+][In^-]}{[H\mathrm{l}n]}$$

Where $[H^+]$, $[In^-]$ and [HIn] are the concentrations of these species in moles

K_{in} = Dissociation constant of the indicator

For acid-base titration the indicator selected should change its colour at the equivalance point or neutralization point of the titration.

Consider the titration of a weak acid like CH_3COOH (0.1 N) against a strong base NaOH (0.1 N) taken in burette. The equivalence point or neutralization point is attained when all the acetic acid is neutralized to CH_3COONa. The pH at this equivalence or neutralization point is 8.7. (Since CH_3COONa is a salt of a weak acid and strong base, therefore its pH remain in alkaline range). Phenolphthalein is the correct indicator for this titration since it is colourless below pH 8.5 and appears pink at and above 8.5. Methyl red should not be used in this titration since it is red below pH 6.3 and yellow at and above 6.3. Therefore, if this indicator is used, colour change of the indicator occurs before the neutralization point indicating the wrong arrival of the end point.

Methyl red can be used for the titration between a strong acid like HCl and a strong base like NaOH. When a 0.1 N HCl is titrated against a 0.1 N NaOH taken in burette, the equivalence point is attained with neutralization of all the HCl forming NaCl. The pH at the equivalence or neutralization point is 7.0 since NaCl is the salt of a strong acid and strong base. The pH of HCl is below 6.3 just before the neutralization point. With addition of only one or two drops of NaOH the pH suddenly rises to 7.0 and methyl red appears yellow giving a correct indication of the arrival of the end point. Use of phenolphthalein indicator in the titration between a strong acid and strong base is a wrong selection.

Redox Indicators

A redox indicator undergoes oxidation and reduction reactions. It exhibits one colour in the reduced state and a different colour in the oxidized state. These indicators are used in titrations involving oxidation-reduction reactions i.e when titration is carried out between an oxidizing agent and a reducing agent. The best known examples of redox indicators are diphenylamine and ferroin. The basic principle of selection of redox indictor is that the standard reduction potential of the indicator should be higher than the standard reduction potential of the reducing agent but less than the standard reduction potential of the oxidizing agent.

Metal ion indicators

Metal ion indicators are used in titrations involving complex formations. The colour of the indicator is changed when it forms complex with a metallic cation.

A well known example is Eriochrome-Black-T (EBT). This compound is sodium 1-(1-hydroxy-2-naphthylazo) – 6 – nitro – 2 napthol – 4 – sulphonate (I). The indicator is used in determining Ca^{++} and Mg^{++} by EDTA (Ethylene diamine tetraacetic acid) titration.

The indicator is blue between the pH 7 and 11 due to dissociation and neutralization of proton from one of the phenolic group. But the indicator when forms complex with Ca^{++} or Mg^{++} appears red. The basic principle of selecting the metallic indicator is that the stability constant of the indicator with the metallic cation to be determined should be less than the stability constant of the complexing reagent (used in the titration) with that metallic cation. In the estimation of Ca and Mg by EDTA titration using EBT indicator the K_{st} (stability constant) of Ca-EDTA or Mg-EDTA is greater than the K_{st} of Ca-EBT or Mg-EBT.

1.6 Determination of Equivalent Mass

Atomic weight

Atomic weight of an element is defined as the relative average weight of its atom as compared to the weight of an atom of C^{12} (stable isotope) taken as 12 on the a.m.u. (atomic mass unit) scale.

1 a.m.u. = 1.66×10^{-24} g = 1.66×10^{-27} kg

Weight of a hydrogen atom = 1 a.m.u.

Weight of an oxygen atom = 16 a.m.u.

Weight of a calcium atom = 40 a.m.u.

Gram atomic weight

Atomic weight of an element expressed in grams is the gram atomic weight.

One gram atomic weight of C = 12 g

One gram atomic weight of N = 14 g

One gram atomic weight of Ca = 40 g

One gram atomic weight of an element consists of Avogadro's number (6.023×10^{23}) of atoms.

Molecular weight

Molecular weight of a substance is defined as the relative average weight of its one molecule as compared to the weight of a carbon atom taken as 12 on the a.m.u. scale.Molecular weight of a substance is the sum of the atomic weights of the constituent atoms in its molecule.

Molecular weight of N_2 = 2×14 = 28 a.m.u.

Molecular weight of $Ca(H_2PO_4)_2$

= (At wt. of Ca) + (4 × At wt. of H) + (2 × At wt. of P) + (8 × At wt. of O)

= 40 + (4×1) + (2×31) + (8×16) a.m.u. = 234 a.m.u.

Gram molecular weight

Molecular weight of an element or compound expressed in grams is the gram molecular weight.

One gram molecular weight of N_2 = 28 g

One gram molecular weight of NaCl = 58.5 g

One gram molecular weight of $Ca(H_2PO_4)_2$ = 234 g

One gram molecular weight of an element or compound consists of Avogadro's number (6.023×10^{23}) of molecules.

Equivalent weight

It is the number of parts by weight of an element or compound that can contain or replace one part by weight of H or equivalent amount of any active element i.e. 8 parts by weight of oxygen, 35.5 parts by weight of chlorine or 23 parts by weight of sodium and so on.

Gram equivalent weight

Equivalent weight expressed in grams is the gram equivalent weight.

One gram equivalent weight of NaCl is 58.5 g

a. Equivalent weight of acids

An acid is a proton donor.

$$\text{Equivalent weight of an acid} = \frac{\textit{Molecular}\ \text{weight of the acid}}{\text{No. of protons it furnishes}}$$

$$= \frac{\textit{Molecular}\ \text{weight of the acid}}{\text{No. of replaceable } H^+}$$

Thus equivalent weight of HCl, HNO_3 are equal to their molecular weights since these acids contain one replaceable H^+ only.

$$\text{Equivalent weight of } H_2SO_4 = \frac{Mol.\ wt.\ of\ H_2SO_4}{2} \text{ (No. of replaceable } H^+ = 2)$$

$$\text{Equivalent weight of } H_3PO_4 = \frac{Mol.\ \text{wt. of } H_3PO_4}{3} \text{ (No. of replaceable } H^+ = 3)$$

$$\text{Equivalent weight of organic acids} = \frac{Molecular\ \text{weight}}{\text{No. of COOH groups}}$$

(since each COOH group contains one replaceable H^+.)

Thus eq. wt. of CH_3COOH = Mol. Wt. of CH_3COOH

Although CH_3COOH contains 4 hydrogen but the number of replaceable H is one to contribute to acidity.

b. Equivalent weight of bases

A base is a proton acceptor.

$$\text{Equivalent weight of a base} = \frac{Mol.\ \text{wt. of the base}}{\text{No. of protons it accept}}$$

Equivalent weight of NaOH = Molecular weight of NaOH (Because NaOH contains one OH^- which accept one proton)

$$\text{Equivalent weight of } Ca\,(OH)_2 = \frac{Mol.\ \text{wt.of } Ca(OH)_2}{2}$$

$$\text{Eq. wt of } Al\,(OH)_3 = \frac{Mol.\ \text{wt.of } Al\,(OH)_3}{3}$$

Equivalent weight of NH_3 = Molecular wt. of NH_3

Because NH_3 accepts one proton by the reaction $NH_3 + H^+ \rightarrow NH_4^+$

c. Equivalent weight of a salt

$$= \frac{\text{Mol. wt.}}{\text{No. of positive or negative charges in the compound}}$$

Equivalent wt of NaCl = Mol. wt. of NaCl

(since in NaCl there is one positive charge and one negative charge)

$$\text{Equivalent weight of } CaCl_2 = \frac{\text{Mol. wt. of } CaCl_2}{2}$$

$$\text{Equivalent weight of } Al_2(SO_4)_3 = \frac{\text{Mol. wt. of } Al_2(SO_4)_3}{6}$$

(since $Al_2(SO_4)_3$ have 6 + and 6 – charges)

$$\text{Equivalent weight of sodium oxalate as a salt} = \frac{\text{Mol. wt. of sodium oxalate}}{2}$$

Sodium oxalate is the salt of NaOH, and oxalic acid

$$\begin{array}{l} COOH \\ \quad | \\ COOH \end{array} + 2\,NaOH \rightleftharpoons \begin{array}{l} COONa \\ \quad | \\ COONa \end{array} + 2\,H_2O$$

(Oxalic acid) (Sodium oxacate)

d. Equivalent weight of an oxidiziong agent

An oxidizing agent is one which captures or gains electrons from another element or compound to oxidize that element or compound and itself is reduced.

$$\text{Equivalent weight of the oxidizing agent} = \frac{\text{Mol. weight of the oxidizing agent}}{\text{No. of electrons it gains}}$$

Or

$$\frac{\text{Mol. wt. of the oxidizing agent}}{\text{Decrease in the oxidation number of the element or the element in a compound whose oxidation number decreases after the redox reaction}}$$

Examples of oxidizing agents are $K_2Cr_2O_7$, $KMnO_4$

Equivalent weight of $K_2Cr_2O_7$

$$K_2Cr_2O_7 + 6\,FeSO_4 + 7\,H_2SO_4 \rightleftharpoons K_2SO_4 + Cr_2(SO_4)_3 + 3\,Fe_2(SO_4)_3 + 7\,H_2O$$

The ionic equation of this reaction is represented as

$$Cr_2O_7^{=} + 14\,H^+ + 6\,Fe^{++} + \rightleftharpoons 2\,Cr^{+++} + 6\,Fe^{+++} + 7\,H_2O$$

This is a redox reaction in which $FeSO_4$ is oxidized by losing electrons to $K_2Cr_2O_7$. Each Fe^{++} loses one electron. Six Fe^{++} loses 6 electrons and those 6 electrons are gained by $K_2Cr_2O_7$. Therefore eq. wt. of $K_2Cr_2O_7 = \frac{\text{Its mol. wt.}}{6}$

In $K_2Cr_2O_7$ the oxidation number of Cr only changes after reaction with $FeSO_4$

Oxidation number of the 2 Cr in $K_2Cr_2O_7$ is 12.

After $K_2Cr_2O_7$ is reduced by $FeSO_4$, the Cr which is present in atomic form in $K_2Cr_2O_7$ changes to the ionic form of Cr^{+++} with formation of $Cr_2(SO_4)_3$ Oxidation no. of the two Cr^{+++} in $Cr_2(SO_4)_3$ is 6

Oxidation no. of the 2Cr in $K_2Cr_2O_7$ before reduction = 12

Oxidation no. of the 2Cr after reduction of $K_2Cr_2O_7$ = 6

∴ Decrease in oxidation number of the 2Cr after reduction = 6

$$\therefore \text{Equivalent wt. of } K_2Cr_2O_7 = \frac{Mol.\text{ wt. of } K_2Cr_2O_7}{6}$$

Equivalent weight of $\boldsymbol{KMnO_4}$

$KMnO_4$ is a mild oxidizing agent. In acid medium $KMnO_4$ gains 5 electrons, whereas in alkaline medium $KMnO_4$ gains 3 electrons. Therefore, equivalent weight of $KMnO_4$ in acid medium is its molecular weight/5, whereas in alkaline medium its eq. wt. is its mol. wt./3.

Calculation of the eq. wt. of $\boldsymbol{KMnO_4}$ ***in acidic medium***

$KMnO_4$ oxidizes sodium oxalate $[(COONa)_2]$ in presence of H_2SO_4 as follows.

$2KMnO_4 + 5(COONa)_2 + 8\ H_2SO_4 \rightarrow K_2SO_4 + 2MnSO_4 + 5\ Na_2SO_4 + 10CO_2 + 8H_2O$

Each molecule of Na oxalate loses 2 electrons.

Five molecules of Na oxalate lose 10 electrons.

These 10 electrons are gained by the 2.

Each therefore, gains 5 electrons

$$\text{Therefore, equivalent weight of } KMnO_4 = \frac{\text{Its mol. wt.}}{5}$$

Calculation of equivalent weight of $\boldsymbol{KMnO_4}$ ***from the decrease in the oxidation number of Mn***

Oxidation number of Mn in MnO_4 is +7

Oxidation no of Mn after reduction is +2

∴ Decrease in oxidation number of Mn after its reduction is 7-2=5

$$\therefore \text{Equivalent weight of } KMnO_4 \text{ in acid medium} = \frac{\text{Its mol. wt.}}{5}$$

Calculation of the equivalent weight of $KMnO_4$ in alkaline medium

$KMnO_4$ oxidizes amino acid in presence of NaOH as follows.

$$2KMnO_4 + 5H_2O \xrightarrow{NaOH} 2KOH + 2\,Mn(OH)_4 + 3O \quad (1)$$

$$3\,RCH\,NH_2\,COOH + 3O \rightarrow 3\,RCOCOOH + 3\,NH_3 \quad (2)$$

$$2KMnO_4 + 3\,RCH\,NH_2\,COOH + 5H_2O \xrightarrow{NaOH} 2KOH + 2\,Mn(OH)_4 + 3\,RCOCOOH + 3\,NH_3$$ (Adding eqn. 1 and 2)

Oxidation no. of Mn in $KMnO_4$ is + 7

After reduction of $KMnO_4$ the oxidation no. of Mn is +4

Decrease in oxidation no. of Mn after reduction of $KMnO_4$ is +3

Therefore, equivalent weight of $KMnO_4$ in alkaline medium

$$= \frac{\text{Mol. wt. of } KMnO_4}{3}$$

In the above reaction there is no transfer of electrons from the amino acid to $KMnO_4$. In other words, amino acid does not lose any electron to $KMnO_4$ to get oxidised. It is oxidised with removal of H and addition of O. The other definition of oxidation is the removal of H and addition of oxygen. Therefore in this type of reaction equivalent weight of the oxidizing agent is only calculated from the decrease in the oxidation no. of the element in the oxidizing agent after reduction.

Equivalent weight of reducing agent

A reducing agent is one that furnishes electrons to reduce another species and, therefore, itself is oxidized by loosing the electrons.

Equivalent weight of the reducing agent

$$= \frac{\text{Its molecular weight}}{\text{No. of electrons it furnishes}}$$

or, $$\frac{\textit{Its}\text{ molecular weight}}{\textit{Increase in}\text{ the oxidation no. of the element or the element in the compound whose oxidation no. increases after the redox reaction.}}$$

Example of the reducing agents is $FeSO_4$, Na oxalate.

Equivalent weight of $FeSO_4$

See the redox reaction between $K_2Cr_2O_7$ and $FeSO_4$ at previous page.

One molecule of $K_2Cr_2O_7$ is reduced by gaining 6 electrons from 6 molecules of $FeSO_4$. Each molecule of $FeSO_4$ furnishes one electron by the reaction $Fe^{++} \rightarrow Fe^{+++} + e^-$.

Therefore, eq. wt. of Fe SO_4 = Its molecular weight.

Calculation of the eq. wt. of $FeSO_4$ from the increase in the oxidation no. of Fe

Oxidation no. of $Fe^{++} = +2$

Oxidation no. of $Fe^{+++} = +3$

∴ Increase in oxidation no. of Fe after the reaction = +1.

So, its eq.wt. = mol. wt.

Equivalent weight of Na oxalate:

See the redox reaction between $KMnO_4$ and Na oxalate in page 12.

Each molecule of Na oxalate loses 2 electrons to $KMnO_4$.

$$\therefore \text{ Eq.wt. of Na oxalate} = \frac{\text{Its molecular wt.}}{2}$$

Equivalent wt. of Na oxalate can also be calculated from the increase in the oxidation no. of carbon after the redox reaction (oxidation no of Na and O remain same after the redox reaction)

$$\begin{matrix} COO^- \\ COO^- \end{matrix} \longrightarrow 2CO_2 + 2e^-$$

Let the oxidation number of C = x

Oxidation no. of O = - 2

$$\text{Oxidation no. of } \begin{matrix} COO^- \\ COO^- \end{matrix} = -2$$

Therefore 2 × (oxidation number of C) + 4 × (oxidation number of O) = –2

⇒ 2 × (oxidation number of C) + 4 × (–2) = – 2

⇒ 2 × (oxidation number of C) = 6

∴ Oxidation no. of C = 3.

Oxidation no. of 2C = 6 (in oxalate ion before reaction)

Oxidation no. of 2C = 8 (when CO_2 is formed after reaction)

∴ Increase in oxidation number of 2C after reaction = 8 – 6 = 2

∴ Equivalent weight of Na-oxalate as the reducing agent = $\frac{\textit{Its mol. wt.}}{2}$

Let us take the example of a redox reaction given below :

$$2\,Cr(OH)_3 + 4OH^- + 1O_3^- \rightarrow 2\,CrO_4^{2-} + I^- + 5H_2O$$

This reaction occurs in presence of OH^- i.e. in an alkaline medium. $Cr(OH)_3$ is oxidized to due to CrO_4^{2-} removal of all the 3H and addition of one oxygen. Here, $Cr(OH)_3$ is the reducing agent and IO_3^- is the oxidizing agent. In this reaction the equivalent weights of $Cr(OH)_3$ and IO_3^- cannot be determined from the loss and gain of electrons. To calculate the eq. wt. of $Cr(OH)_3$ and IO_3^-, the change in oxidation no. are to be considered.

Equivalent weight of $Cr(OH)_3$ as the reducing agent

In IO_3^- $Cr(OH)_3$, the oxidation number of Cr only changes after reaction.

Oxidation no. of Cr in $Cr(OH)_3$ before reaction is +3,

Oxidation no. of Cr in CrO_4^{2-} after reaction is +6,

Increase in oxidation number of Cr after reaction is +3.

Equivalent weight of $Cr(OH)_3$ as the reducing agent = $\frac{\textit{Its mol. wt.}}{3}$

Equivalent weight of IO_3^- as the oxidizing agent

In, IO_3^- oxidation number of I only changes after reaction

Oxidation no. of I before reaction is +5,

Oxidation no. of I after reaction is –1,

∴ Decrease in oxidation number of I after reaction = +6

∴ Equivalent weight of as the oxidizing agent = $\frac{\textit{Ionic wt. of } IO_3^-}{6}$

Equivalent weight of a complex forming or chelating compound

Equivalent weight of a complexing or chelating compound is that part by weight of the compound that can complex one gram equivalent of the cation.

$$\text{Equivalent weight of a cation} = \frac{\textit{Ionic}\text{ weight of the cation}}{\text{no. of charges carried by the cation}}$$

Example : Ethylene diamine tetracetic acid (EDTA)

Ca^{++} + EDTA = Ca – EDTA

∴ Equivalent weight of EDTA

$$= \frac{\text{Mol. wt. of EDTA}}{2} \text{ (1mol of } Ca^{++} = 2 \text{ geq of } Ca^{++})$$

Since EDTA is insoluble, the disodium salt of EDTA (represented as Na_2H_2Y) is used for analysis.

$Na_2H_2Y + CaCl_2 = CaH_2Y + 2\,NaCl$

The equivalent weight of Na_2H_2Y is that part by weight of Na_2H_2Y that reacts with one geq of $CaCl_2$ (one g equivalent of $CaCl_2$ contains one g equivalent of Ca^{++}). In the above reaction 1 mole of Na_2H_2Y reacts with 1 mole of $CaCl_2$ or, 1 mole of Na_2H_2Y reacts with 2 g eq. of $CaCl_2$

$$\therefore \text{ Equivalent weight of } Na_2H_2Y = \frac{\text{Mol. wt. of } Na_2H_2Y}{2}$$

1.7 Expression of Concentration of Solutions

Standard solutions are solutions of known strength. Strength of a solution is expressed in molarity, normality, percentage and parts per million.

a) Molarity or molar concentration

Molarity of a solution is the no. of gram moles of the solute dissolved in 1 litre of the solution. A 'x' molar solution of a substance contains x gram mole of that substance in 1 litre of the solution.

b) Normality or normal concentration

Normality of a solution is the no. of gram equivalents of the solute dissolved in 1 litre of the solution.

A 'x' normal solution of a substance contains x gram equivalent of that substance in 1 litre of the solution.

c) Percentage concentration

The concentration of the solute in a solution may be expressed in percentage as follows.

i) weight per cent (w/w) $= \frac{\textit{weight of the solute}}{\text{weight of the solution}} \times 100$ (same unit of weight is to be used for both solute and solution)

ii) volume per cent (v/v) $= \frac{\text{volume of the solute}}{\text{volume of the solution}} \times 100$ (same unit of volume is to be used for both solute and solution)

$$\text{weight - volume percent (w/v)} = \frac{\textit{weight of the solute}(g)}{\text{volume of the solution (mL)}} \times 100$$

d) Parts per million (ppm)

It is the no. of the parts of the solute dissolved in 10^6 parts of solution. It may be expressed in weight / weight (w/w), weight / volume (w/v) or volume / volume (v/v) basis.

$$\text{ppm (w/w)} = \frac{\text{wt. of the solute}}{\text{wt. of the solution}} \times 10^6$$

$$\text{ppm (w/v)} = \frac{\text{wt. of the solute (gm)}}{\text{volume of the solution (mL)}} \times 10^6$$

$$\text{ppm (v/v)} = \frac{\text{volume of the solute}}{\text{volume of the solution}} \times 10^6$$

The concentration of a solute in very dilute solution can be conveniently expressed in ppm. A one ppm solution of P (w/v) contains 1 g of P in 10^6 mL of the solution or 1 mg of P in 10^3 mL or 1 litre of the solution.

2

Principles of Analytical Instruments

Advances in analytical chemistry and instrumental analytical chemistry in the past few decades have significantly improved, making elemental determinations in soil extracts and water samples easier, faster and less challenging for the analyst. For most of the elemental analysis procedures, the more traditional wet chemistry procedures have been replaced by various analytical instrumental procedures that employ either emission or absorption spectrophotometry. The elemental concentration in prepared soil extracts and water samples can be determined by a number of instrumental analytical procedures.

The trend today is toward greater analytical sophistication, to multielement computer controlled analytical instrumentation, resulting in a less ended understanding of the analytical principles involved, and "black box" concepts of instrument calibration, maintenance, and operation. Preparation of the analyte and suitable adaptation for a method of analysis requires an understanding, by the analyst, of the principle of the method as well as its requirements and limitations. Adequate testing is usually required before putting the method to use, following procedures such as those that have been adapted by the Association of Official Analytical Chemists (AOAC).

It is not possible to cover in adequate detail every method of analysis suited for the elemental concentration determination in soil extracts and water samples. Therefore, the objective here is to provide sufficient background information about the instrumental principles to guide the reader.

2.1 pH Meter

A pH meter measures essentially the electro-chemical potential between a known liquid inside the glass electrode (membrane) and an unknown liquid outside. Because the thin glass bulb allows mainly the agile and small hydrogen ions to interact with the glass, and the glass electrode measures the electro-chemical potential of hydrogen ions or the *potential of hydrogen*. To complete the electrical circuit, also a reference electrode is needed. The instrument does not measure a current but only an electrical voltage. It measures the electrical potential between the mercuric chloride of the reference electrode and its potassium chloride liquid, the unknown liquid, the solution inside the glass electrode, and the potential between the solution and the silver electrode. The electrodes are given in the respective figures 2.1. Most modern pH meters also have a thermistor temperature probe which allows for automatic temperature correction, since pH varies somewhat with temperature.

pH Cell

The pH determination with the help of pH meter is based on the measurement of the electromotive force (EMF) of a pH cell consisting of an electrode sensitive/responsive to H^+ ions and a reference electrode. The pH response is confined to the area of the glass membrane. Both the inner and outer surface of the bulb of the glass electrode is sensitive to H^+ ion concentrations. The inside of the bulb is filled with an electrolyte of high buffering capacity such as a 0.025 M solution of HCl. However, the reference calomel electrode consists of a paste of Hg along with Hg_2Cl_2 filled in the inside tube (or Ag and AgCl is used in some electrodes) which is connected to the KCl solution outside the tube.

The pH cell can be represented as;

H + sensitive Electrode	References Buffer or Test Solution	Salt Bridge	Reference Electrode

The chemical equilibrium;

$AgCl\ (s) + e^- = Ag^+ + Cl^-$

$K_{sp} = 1.8 \times 10^{-10}$

Nernst equation

$E = 0.799 + 0.0591\ log\ K_{sp} - 0.0591\ \log\ [Cl^-]$

When the glass electrode bulb (made up of lithium silicates) of the pH meter is dipped in a solution having H^+ ions concentrations different from that inside the electrode, an electrical potential gradient develops across the membrane and this potential is proportional to the difference in the pH between two sides. Since the pH of the inner side remains unchanged, the difference denotes the H^+ ions concentrations of the outer solution.

Soil pH controls the nutrient availability, toxicity and microbial activity in soil. Soil pH is the negative logarithm of H^+ activity in the soil solution. There is difference between activity and concentration. Activity and concentration are related by the mathematical expression;

$$\frac{a}{c} = f$$

Where;

a = activity of a species (moles L^{-1})

c = concentration of that species (moles L^{-1})

f = activity coefficient

Activity coefficient of an ionic species is the degrees of freedom of that species to participate in an equilibrium reaction or to maintain its charge under an ionic environment.

Activity coefficient of an ionic species is determined by Dubye – Huckel equation (1) taking into consideration of the ionic strength of the solution, charge of the ionic species and effective diameter of the species in hydrated state. The maximum value of activity coefficient is 1. As the concentration of an ionic species increases the activity coefficient decreases and vice versa.

$$\log \gamma_i = \frac{AZ_i^2 \text{-} \sqrt{I}}{1 + Bd\sqrt{I}} \qquad \text{eq (1)}$$

Where, $I = \frac{1}{2}\sum C_i Z_i^2$

Ci = Conc. of the ion (moles/L)

Zi = Valence of the ion

A&B = Constants

d = diameter of ion

Comparing a 0.1 M HCl and 0.01 M HCl, activity coefficient of H^+ is greater in the latter than in the former. Because of the less ionic population of H^+ and Cl^- in 0.01 M HCl, the H^+ remains at a greater distance from the surrounding Cl^- ions. Therefore, the electrostatic force of attraction between H^+ and Cl^- becomes less. So less of the + charge of H^+ is neutralized and more remains unneutralized. In other words the H^+ in 0.01 M HCl has a greater freedom to retain its positive charge than in 0.1 M HCl.

All types of equilibrium reactions are affected by the activity of the ionic species rather than the concentration of the ionic species. Let us take the example of acetic acid;

$$CH_3\,COOH \rightleftharpoons CH_3\,COO^- + H^+$$

The dissocitation of CH_3COOH increases with addition of NaCl. After addition of NaCl the CH_3COO^- is surrounded by Na^+ and H^+ is surrounded by Cl^-. Therefore, their degrees of freedom to get combined with each other to form CH_3COOH decreases. In other words, their activity coefficient decreases and consequently their activity also decreases. Therefore, the rate of the reverse reaction decreases and the equilibrium proceeds in the forward direction. In other words more of the CH_3COOH dissociates to maintain the equilibrium. In the similar process, the solubility of a sparingly soluble salt increases by adding a salt consisting of cation and anion different from those in the sparingly soluble salt.

pH Scale

The pH scale of water is used to measure the pH of the soil solution since water is the solvent in the soil solution. Water weakly dissociates to H^+ and OH^-.

$$H_2O \rightleftharpoons H^+ + OH^-$$

At equilibrium, $k_f a_{H_2O} = K_r a_{H^+} a_{OH^-}$

Where, a_{H_2O}, $a_{H^+} a_{OH^-}$ are the activities of H_2O, H^+ and OH^-, respectively.

k_f = Rate constant for the forward reaction

k_r = Rate constant for the reverse reaction

$\Rightarrow K = k_r a_{H^+} a_{OH^-}$ (Since water is a pure solvent, $k_f a_{H_2O}$ is constant)

$$\Rightarrow a_{H^+} a_{OH^-} = \frac{K}{k_r} = K_w$$

(Since K and k_r are constants, K_w is also a constant called the dissociation constant of water. K_w is also called the ionic product of water)

$\Rightarrow a_{H^+} a_{OH^-} = 10^{-14}$ at 25°C ($\because K_w$ at 25°C $= 10^{-14}$)

$\Rightarrow -\log a_{H^+} + \left(-\log a_{OH^-}\right) = 14$ (at 25°C)

$\Rightarrow$ pH + pOH = 14

Therefore, an inverse relationship exists between pH and pOH. If the pH of a solution is 5, then pOH of that solution is 9.

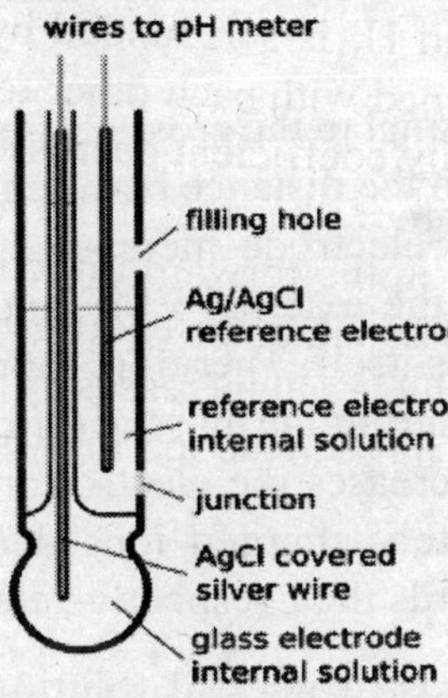

(a) Glass Electrode

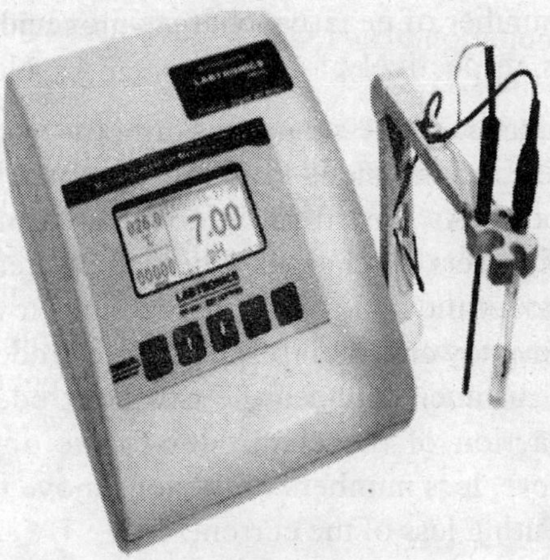

(b) pH Meter

Fig. 2.1: Glass electrode and pH meter

2.2 Conductivity Meter

Conductance of a solution is measured by a conductivity meter which consists of a conductivity cell that is dipped into the solution whose conductance is to be measured. The conductivity cell contains two platinum electrodes of rectangular size spaced at a specific distance. When the electrodes are connected to an external source of electromotive force, current flows through the solution or in other words the solution conducts electricity.

$$\text{Conductance} = \frac{Ampere}{Volt}$$

The unit of conductance is mhos. The current strength is expressed in ampere. One ampere is the flow of 1 coulomb of charge per second. The charge of one gramequivalent of a cation or an anion is 96,500 coulombs. Therefore 1 coulomb of charge is carried by $\frac{1}{96,500} = 1.04\times10^{-5}$ geq of cations or anions. If the conductance of a solution is 1 mhos that means 1.04×10^{-5} gramequivalent of cations and anions move per second towards the respective electrodes under an applied potential difference of one volt. Therefore, if the conductance of a solution is 1 mmhos, then 1.04×10^{-5} milliequivalent of cations and anions move per second towards the respective electrodes under the applied potential difference of one volt. Since 1 geq of a cation or anion carries 6.023×10^{23} no. of positive or negative charges (Avogadro's number), therefore one milliequivalent of the cation or anion carries 6.023×10^{20} number of positive or negative charges. This shows when the conductance of a solution is 1 mmhos, then $(1.04\times10^{-5}) \times 6.023 \times10^{20} = 6.264 \times10^{15}$ no. of positive charges and the same number of negative charges move per second towards the cathodes and anodes, respectively.

Conductance of the solution is directly proportional to the cross sectional area of the electrodes and is inversely proportional to the distance between the two electrodes. As the cross sectional area of the electrode increases, it bears more numbers of charges when connected to the external EMF and attract more no. of the oppositely charged ions towards itself. Therefore more of the ions move towards their respective electrodes contributing more of the currents. As the distance between the two electrodes increases the electrostatic force of attraction of the electrodes to the oppositely charged ions decreases. Therefore, less numbers of the ions move towards their respective electrodes contributing less of the current.

$$C \propto a$$

$$C \propto \frac{1}{\ell}$$

C = conductance of the solution in mmhos

ℓ = distance between the two electrodes (cm)

$$\Rightarrow C \propto \frac{a}{\ell}$$

a = cross sectional area of the electrode (cm^2)

$$\Rightarrow C = k \times \frac{a}{\ell}$$

k is the proportionality constant called specific conductance or electrical conductivity of the solution (mmhos cm^{-1}).

$$\Rightarrow k = C \times \frac{\ell}{a}$$

ℓ/a = cell constant (cm^{-1})

To evaluate k, k = C when a =1 cm^2 and ℓ= 1 cm. Therefore, specific conductance or electrical conductivity of a solution is its conductance when the cross sectional areas of the electrodes are 1 cm^2 and the electrodes are spaced at 1 cm apart. In other words, specific conductance is the conductance of each one cc of the solution. Conductance of a solution may vary from instrument to instrument because of the variation in 'a' and 'ℓ' of the electrodes. But specific conductance is constant at a particular temperature irrespective of the values of 'a' and 'ℓ' of the electrodes in the instrument.

Conductance of a solution increases with increase in temperature due to increase in the ionic mobility. Ionic mobility is the distance travelled by an ion per second towards the respective electrode under unit potential gradient i.e. a potential gradient of one volt per cm.

$$\text{Potential gradient} = \frac{\text{Potential differnence applied between the two electrodes (volt)}}{\text{Distance between the two electrodes (cm)}}$$

The increase in ionic mobility with increase in temperature is possibly due to increase in the intermolecular distance among the water molecules with rise in temperature. The ions, therefore, get a greater free path for movement towards their respective electrodes and can cover a greater distance per second. Therefore more numbers of ions move towards their electrodes contributing more of currents. The reverse occurs with decrease in temperature. Therefore for uniform comparison, electrical conductivities of the solutions are expressed at 25°C. Since above 25°C, the EC value increases, therefore it is multiplied by a factor <1.0 to convert it at 25°C. As the temperature increases more and more above 25°C, EC increases more and more and the value of this factor also become less and less than 1. Since the EC value decreases below 25°C therefore it is multiplied by a factor > 1.0 to convert the value at 25°C. As the temperature decreases below 25°C, the value of this factor increases above 1.0.

$$\text{Cell constant of the instrument } (\ell/a) = \frac{Csp \text{ of 0.02 M KCl at 25°C}}{Cm \text{ of 0.02 M KCl at 25°C}}$$

Csp = Specific conductance (mhos/cm^{-1})

Cm = Measured conductance (mhos)

Csp of 0.02 M KCl at 25°C = 2.768 mhos/cm^{-1}

Cm of 0.02 M KCl at 25°C =

Cm of 0.02 M KCl at the measured temperature × Temp. factor

Csp of the soil solution at the measured temperature = *Cm* of the soil solution × ℓ/a.

CSP of the soil solution at 25°C = Csp of the soil solution at the measured temperature × temperature factor.

N.B. Electrical conductivity (EC) of the soil refers to the specific conductance of the soil solution. EC decreases as the soil : water extraction ratio increases and vice versa. Electrical conductivity of a soil at 1:1 soil : water extraction ratio is more than that at 1:2 soil : water extraction ratio. Therefore, while writing the EC of a soil the soil : water extraction ratio should be specified (in brackets by the side of EC). For example = EC (1:2) = 0.181 dS m^{-1} in Bhubaneswar soil.

Concentration of salts in the soil water extract or irrigation water (meq per litre) = 12.5 × L mmhos cm^{-1}

Where L mmhos cm^{-1} is the specific conductance of the soil water extract at 25ºC.

Conc. of salts in the soil solution (ppm) = 640 × L mmhos cm^{-1}

% salt in soil = 0.064 × L mmhos cm^{-1} × water : soil of extraction

Csp of the saturation extract of soil

$$= \text{Csp of the soil (1:2)} \times \frac{200}{\%\ \textit{water}\ \text{in soil at saturation}}$$

Conversion of mmhos cm^{-1} to dS m^{-1}

1 Siemen = 1 mhos

Let the conductance = C mmhos = C milliSiemen = $\frac{C}{100}$ dS

Let the cross sectional area of the electrode = a cm^2 = a × 10^{-4} m^2

Let the distance between the 2 electrodes = ℓ cm = ℓ × 10^{-2} m

$$\therefore \text{Cell constant } \frac{\ell}{a} cm^{-1} = \frac{\ell \times 10^{-2}}{a \times 10^{-4}} = \frac{\ell}{a} \times 10^{2} m^{-1}$$

$$\text{Specific conductance (mmhos cm}^{-1}) = \text{C mmhos } \frac{\ell}{a} cm^{-1} \times = \text{C} \times \frac{\ell}{a} \text{ mmhos cm}^{-1}$$

$$\text{Specific conductance (dS m}^{-1}) = \frac{C}{100} dS \times \left(\frac{\ell}{a} \times 10^{2} m^{-1} \right) = C \times \frac{\ell}{a} dS \text{ m}^{-1}$$

∴ Specific conductance in mmhos cm^{-1} = Specific conductance in dS m^{-1}.

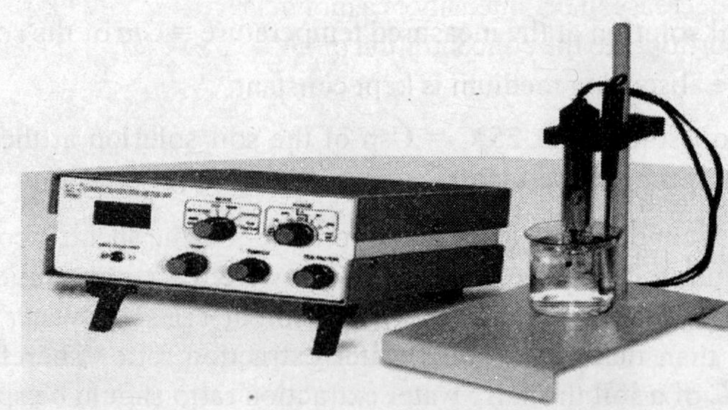

Fig. 2.2: Conductivity meter

Electrical conductivity of the soil is the quantity of electricity in coulombs that passes per second per cm^3 of the soil solution under an applied potential difference of 1 volt. It is written as EC in abbreviated form and is expressed in millimohs cm^{-1} or dS m^{-1} (deciSiemen m^{-1}). Current flows in the soil solution due to presence of soluble salts like Cl^-, SO_4^{2-} and NO_3^- of Na^+, K^+ Ca^{++} and Mg^{++}. These salts fully dissociate in water to their constituent cations and anions and the current is carried by those cations anions in the soil solution. Therefore, electrical conductivity of soil increases with increase in soluble salt content in soil. This experiment is important for identifying the saline and saline-alkali soils. Electrical conductivity of the saturation extract of the saline and saline-alkali soils are > 4 dS m^{-1}. Specific soil management practices are required for reclamation of saline and saline-alkali soils to obtain higher yield potential.

2.3 Spectrophotometer

Spectrophotometer is used to determine the concentration of a substance by the measurement of the relative absorption or transmittance of light in the visible region with respect to a known concentration of the substance. A monochromatic light beam of known intensity is passed through the test solution and the intensity of the transmitted beam is determined with the help of a photoelectric cell.

Instrument used for colorimetric measurement is also called colorimeter. It can be considered as the absorption spectrophotometry in the visible range. Now a days this instrument has been named as spectrophotometer since photoelectric cells are generally used to measure the light transmitted by the solution.

When a beam of radiant energy passes through a medium, the energy of the beam is partly altered by reflection, refraction, diffraction or absorption and the reminder of the energy may be transmitted through the substance. Hence, spectrophotometry or colorimetry is based on Lambert's and Beer's law which states that "the decrease in the intensity of a monochromatic light (exponentially) is directly proportional to the concentration of the solution (arithmetically) if the thickness of the absorbing medium is kept constant".

Hence;

Transmittance $T = I_t/I_0$

Absorbance = $\log 1/T = \log I_0/I_t$

$$\log \frac{I_0}{I_t} = kC$$

Where,

I_0 = the intensity or energy per cm^2 per second before passing through substance.

I_t = the intensity of energy after passing through the medium.

k = constant

C = concentration of the test solution

The light from the source travels through the field lens, entrance slit, and collimating mirror. Collimating mirror allows the light to fall on the grating chamber which disperses the light. The required wavelength of the light is adjusted by adjustable light control (wave length adjustment). The monochromatic light passes through the exit slit, filter and the sample. Some of the light energy is absorbed by the sample and the remaining is transmitted to the photocell which converts light energy into the electrical energy. This transmittance is measured by galvanometer.

Table 2.1: Selection of wavelength range

Wavelength range (nm)	Colour of thesample solution	Complement	Wavelength range (nm)
400 -500	Violet	Yellowish green	560 – 575
450 – 480	Blue	Yellow	575 – 590
480 – 490	Greenish blue	Orange	590 – 625
490 – 500	Bluish green	Red	625 – 750
500 – 560	Green	Purple	——
560 – 575	Yellowish green	Violet	400 -500
575 - 590	Yellow	Blue	450 – 480
590 – 625	Orange	Greenish blue	480 – 490
625 - 750	Red	Bluish green	490 500

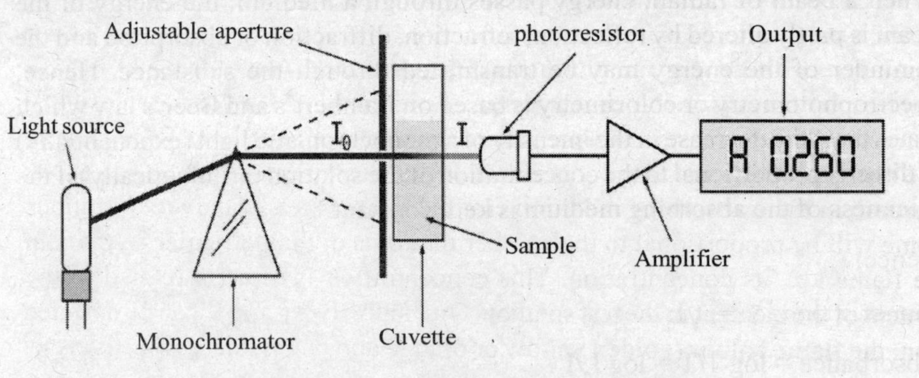

Fig. 2.3: Schematic diagram showing the working principle of spectrophotometer

The major components of colorimeter/spectrophotometer are;

1. Electric bulb – Source of light.
2. Filter – it selects the light radiation of desired wavelength.

3. Sample holder or cuvette. It holds the sample solution.
4. Photocell – Converts light energy into electric energy.
5. Galvanometer – it records the electrical energy as absorbance.

Application: Used in the estimation of available phosphorus and sulphur in soil, different enzymes in soil and determination of concentration of different colour solutions.

Calibration: See the experiment page in the determination of available P & S in soil.

2.4 Flame Photometry

Flame photometer is based on the working principle of atomic emission spectrometry. It is used to estimate the alkali (Na, K) and alkaline earth metals (Ca, Mg) in soil, plant and water samples. The use of flame emission spectroscopy for the estimation of the alkali metals is an important application in routine chemical analysis. For this, low temperature flame photometry provides the most reliable and convenient method. This exploits the fact that compounds of the alkali and alkaline earth metals are thermally dissociated at the temperature of a burner flame and some of the atoms produced are further excited to a higher energy level. When these excited atoms return to the ground state, they emit radiation which, for the alkali and alkaline earth elements, lies mainly in the visible region of the spectrum.

When a solution of a salt is sprayed into a flame the salt breaks into the component atoms due to high temperature. The energy provided by the flame excites the orbital electrons to higher energy levels. When these electrons return to their ground state they emit a characteristic radiation. The flame emission spectroscopy is thus designed to measure the intensity of the characteristic line (resonance line) emitted from the excited atoms or ions (Na^+ - 589 nm, K^+ - 766 nm, Ca^{2+} - 422 nm, Mg^{2+} - 285 nm) of the element to be determined. The increase in the intensity of radiation emitted from a steady and non luminous flame will be proportional to the number of atoms of the particular element in the flame i.e., its concentration. This concentration is directly related to the content of the element in the test solution. Qualitatively Na and K can be detected from the flame colour (golden yellow or orange and pale violet respectively).

The wavelength of light emitted from the flame is characteristic of the particular element. The intensity of this light is proportional to the amount of the salt solution volatilised in the flame at any moment. Hence, a flame photometer essentially consists of the following components.

1. A flame that can be maintained in a constant form and at a constant temperature
2. A means of transporting a homogenous solution in the flame at a steady rate
3. A means of separating the light of the wavelength to be measured from that of extraneous emission with a monochromator or an optical filter
4. A means of measuring the intensity of the radiation emitted from the flame.

The two characteristics of the alkali metals chiefly facilitating their determination by flame photometry are:

(i) Their spectra are excited at a temperature lower than that at which most other elements are excited. This low excitation temperature permits ordinary bottled gas to be used as fuel for the flame.

(ii) Their characteristic wavelengths are widely separated from each other and from those of most other elements. This wide separation of the characteristic wavelengths makes it possible to use simple optical filters rather than the more expensive monochromators.

Table 2.2 Alkali & Alkaline earth metals & their characteristic wavelengths (nm)

Barium	553.6
Calcium	623.0
Cesium	455.6
Lithium	671.0
Potassium	766.0
Rubidium	780.0
Strontium	460.7
Sodium	589.0

Basic Components of Flame Photometer

- Burner
- Mirrors
- Slits
- Monochromators
- Filters
- Detectors

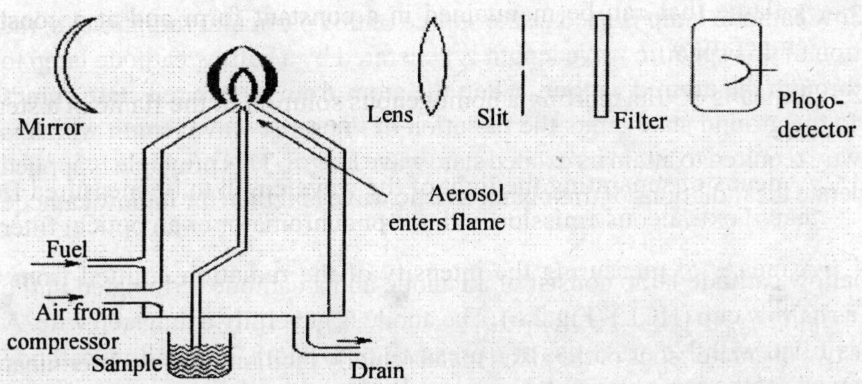

Fig. 2.4: Schematic diagram showing the working principle of flame photometer

Application: Used in the estimation of the concentration of different alkali and alkaline earth metals in the soil water and plant extract.

Calibration: See the experiment page in the determination of available K in soil and Na and K content in water sample.

2.5 Atomic Absorption Spectrophotometer

Principle

In the analysis employing Atomic Absorption Spectrophotometer (AAS) the sample in the form of a homogeneous liquid is aspirated into a flame where "free" atoms of the element to be analysed are created. A light source (hollow cathode lamp) is used to excite the free atoms formed in the flame by the absorption of the electromagnetic radiation. The decrease in energy (absorption) is then measured which follows the Lambert - Beer law, i.e. the absorbance is proportional to the number of free atoms in the ground state.

Atoms of a metallic element in its ground state absorb radiation of a specific wave length to attain the excited state. The valency electron/electrons in the excited state are promoted to higher energy level. When this characteristic radiation is allowed to pass through the vapour of the particular metallic atoms, produced in a flame, a part of this radiation is absorbed and the rest is transmitted. The amount of radiation absorbed is directly proportional to the density or concentration of that atom in the atomic vapour.

By measuring the intensity of the transmitted radiation in presence and absence of atomic vapour, the concentration of the metallic atoms in the soil extract can be determined. This is the basis on which the atomic absorption spectroscopy has been developed.

A hallow cathode lamp is used as the source of this particular radiation i.e. the radiation of this specific wave length is generated by a hallow cathode lamp to pass through the atomic vapour. When the atom from its excited state comes back to the ground state emits the radiation of the same wave length which is otherwise required to attain its excited state wave length. This principle is applied to generate the radiations of this characteristic wave length in the hallow cathode lamp.

The hallow cathode lamp consist of an anode and a cathode terminated in the lamp as hallow cup (HCL) (Fig.2.6). The anode is generally a tungsten wire. A HCL is filled with Ne or Ar gas at a pressure of 1-5 torr. The hallow cathode cup is made of the same element which is to be determined. For example if zinc is to be determined, then the hallow cathode cup should be constructed with Zn metal. If copper is to be determined the hallow cathode cup should constructed with the same metal. It may also be constructed from an inert element into which the desired element or a salt of the desired element is kept. The hallow cathode lamp is filled with an inert gas like neon or argon. The lamp envelope is made of glass. When a dc voltage of 300-500 volt is put across the anode and cathode, the anode then becomes positively charged and the cathode becomes negatively charge. The highly positively charged anode attracts the electrons in the outer most orbits of the neon atoms. The neon atoms loss their electrons in

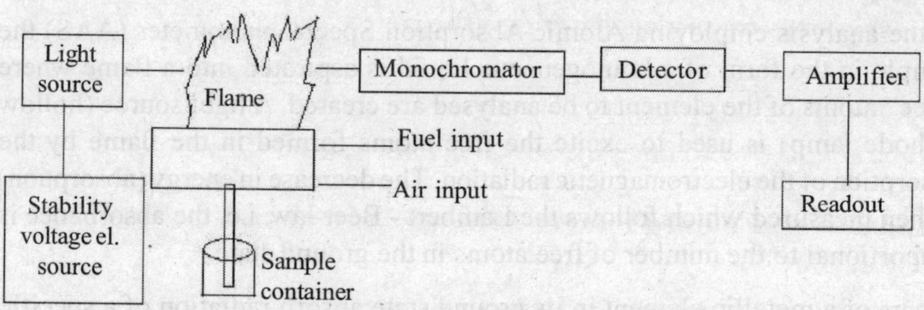

Fig. 2.5: Block diagram of the major components of AAS

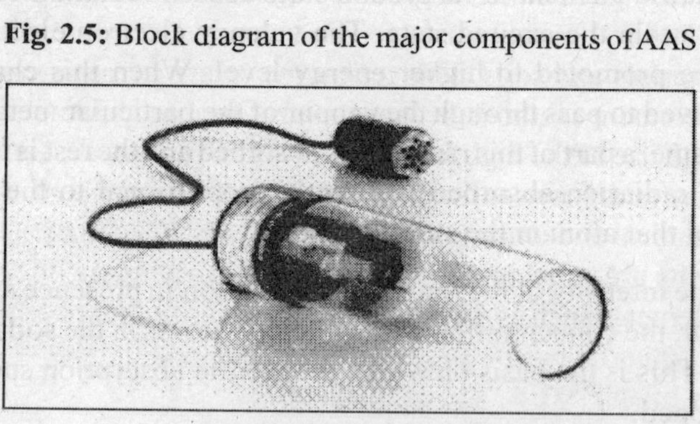

Fig. 2.6: Hallow cathode lamp (HCL)

the outmost shell near the anode and become positively charges. The positively charged neon is then strongly attracted by the cathode and moves to the cathode at a high speed. The neon cation strikes the metallic atom (which is same as the element to be determined) and dislodges the surface metallic atoms in to neon gas. Further collision of these vapourised metallic atoms with the ionized neon atoms causes excitation of these metallic atoms. When these metallic atoms again return to their ground states the excess energy is emitted as electromagnetic radiation whose wave length is same as the wave length required bringing the atoms to its excited state. In addition to the radiation of this wave length other radiation may be emitted from the neon gases. The characteristic radiation of the particular metallic atoms is scanned by a monochromator and is allowed to pass through the metallic vapour in the flame for absorption.

Preparation of standards and sample solutions

- Prepare stock standards in concentrations of 1000 mg/litre from pure metal wire, granules, foil, metal oxides or other suitable primary standard compounds of the elements.
- Prepare a blank sample by using the same reagents as used for standard samples but without the elements intended to be estimated.
- Free the sample solution from interfering elements and suspended solids, which may cause clogging of the nebulizer.
- Adjust both standards and unknowns to a concentration range, which is compatible with the analytical range of the instrument. It should preferably be at least 5 to 10 times the detection limit of the instrument.

2.6 Inductively Coupled Plasma-Atomic Emission Spectroscopy (ICP-AES)

A new analytical technique called inductively coupled plasma atomic emission spectroscopy (ICP-AES) has been used for simultaneous multi element analysis of biological materials and soils. This technique offers advantages over AAS and other multi-element methods because matrix problems are eliminated or minimized through use of the high temperature argon plasma. Apart from multi element capability at all concentration levels, plasmas are noted for relative freedom from chemical and ionization interferences that are common with AAS, and detection limits are equal to or better than AAS, depending on the element to be analysed. Elements such as Al, P, S and B, which are either poorly measured at low concentrations or not possible by AAS, are readily determined with higher sensitivity by ICP-AES.

Principle

ICP-AES is based on the observation of atomic emission spectra when samples in the form of an aerosol, thermally generated vapour or powder are injected into an inductively coupled plasma atomization and excitation source. By definition, plasma refers to a hot gas in which a significant fraction of their atoms or molecules is ionized. Plasmas are electrically conducted and have been referred to as electrical flames, as no combustion takes place. This is because instruments using a plasma source generally use inert argon gas.

The ICP is produced by passing initially ionized argon gas through a quartz torch located within an induction coil (Cu coil) which is connected to a radio frequency (RF) generator. The ratio frequency generator produces 1.5 to 3 KW power at a frequency of 27.1 MHz. An oscillating magnetic field is formed within the quartz torch in response to the radio frequency energy passing through the coil. Electrons and ions passing through the oscillating electromagnetic field flow at high acceleration rates within the quartz torch space. As argon gas enters the magnetic field associated with the induction coil, its atoms collide with the accelerated ions and electrons resulting in the ionization of the argon gas. These collisions give rise to ohmic heating, which produces plasma with temperature ranging from 6000 to 10000 K. The resultant plasma is contained within the torch by means of argon flow.

The method of presenting the sample to the plasma is similar to that used in flame atomic absorption. The liquid sample is aspirated into the plasma through a nebulizer system by using argon carrier gas at a rate of about 1 litre argon/ minute. The prevailing high temperature in the plasma leads to complete vaporiozation, atomisation and excitation of the element to be analysed. The exited neutral atoms or ions of the sample emit radiation of characteristics wavelengths with the intensity of the emitted radiation is measured by the spectrophotometer component of the ICP - AES instrument.

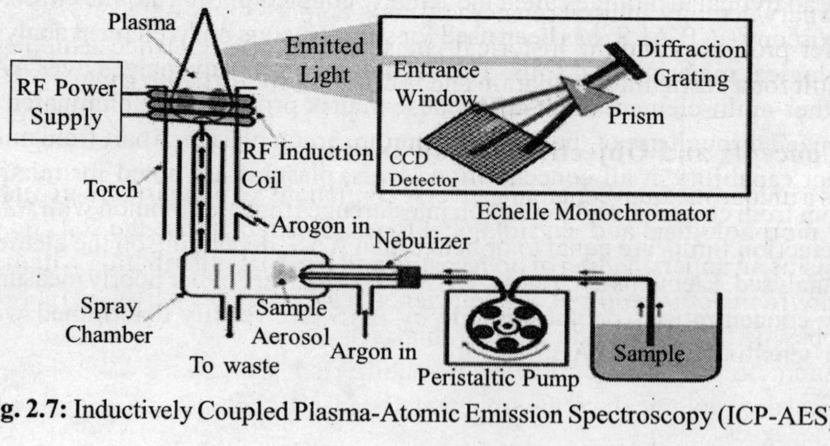

Fig. 2.7: Inductively Coupled Plasma-Atomic Emission Spectroscopy (ICP-AES)

3

Soil Analysis

3.1 Concepts of Soil Sampling

Sampling involves the selection from the total population of a subset of individuals upon which measurements will be made. The measurements made on this subset (or sample) will then be used to estimate the properties (or parameters) of the total population. Sampling is inherent to any field research program in soil science because the measurement of the total population is impossible for any realistic study.

Sampling design involves the selection of the most efficient method for choosing the samples that will be used to estimate the properties of the population. The definition of the population to be sampled is central to the initial formulation of the research study. The sampling design defines how specific elements will be selected from the population, and these sampled elements form the sample population.

Pedogenetic and soil geomorphic studies focus on determining the processes that formed the soil properties or landscapes under study and the environments that controlled the rates of these processes. Pedon-scale studies are closely associated with the development of soil taxonomic systems and focus on vertical, intrapedon processes. Soil geomorphic studies are the interface between quaternary geology and soil science and soil geomorphologists focus on lateral transfer processes and the historical landscape evolution. Hence sampling is difficult for a particular population and is different for different study.

3.2 Concepts and Objectives of Soil Testing

Soil is a unique medium containing diverse community of organisms representing many morphological and physiological types. To characterise the soil and the organisms an understanding of both spatial and temporal distribution within the soil environment is required. The nutrients, water quality and enumeration of microbes or the magnitude of microbial activity represents a point-in-time measurement that is at some dynamic equilibrium governed by the physical, chemical and biological soil environment. Besides variation within the profile, it is reasonable to expect spatial variation in soil mineralogical as well as microbiological properties within sampling area similar to than landscape units. Spatial variation has been documented for a variety of physical and chemical properties including hydraulic conductivity, water content, soil acidity as measured by pH.

Distinct boundaries between soil classification units mapped in a field rarely occur and it is common for units to grade from one to another in a continuum of gradually changing characteristics. Because soil properties vary from location to location and vary with depth or even season of the year, such intrinsic variation must be considered prior to developing sampling points to determine the biochemical properties. Within the highly complex soil environment attempts to study mineralogical and microbiological variations in particular will be difficult. Soil testing gives a general overview on sampling for nutrient analysis. Hence soil testing is the rapid chemical analysis of a soil to estimate soil pH, electrical conductivity, oxidizable carbon, mineralizable N, Brays' and Olsen's P and exchangeable K, etc.

Soil testing identifies various constraints in soil that limit crop production. Fertilizer recommendation and soil management practices to be adopted are based on the soil test report. Soil testing starts with collection of soil samples from the field and ends with interpretation of soil test results leading to necessary recommendations for increasing crop production. To obtain correct soil test values, the soil sample should be a proper representative of the soil of the particular area. Since soil is heterogeneous in nature, specific techniques are followed to collect the soil sample as, it represents the average physical, chemical, biological properties and nutrient status of the soil of that area. One must be well acquainted with these techniques before starting this important work.

Soil testing has long been accepted as a unique tool for rational fertilizer use. In fact, however, soil testing can act as a watchdog to safeguard soil quality as a whole. Thus, the major *objectives of soil testing* are:

i) To assess the soil fertility status and recommend suitable and economic nutrient doses through chemical fertilizers and organic manure for different corps and cropping systems.

ii) To identify the type and degree of degradation problems/ abnormalities like soil acidity, salinity and sodicity etc, and to suggest effective remedial measures.

iii) To generate data for compilation of soil fertility maps.

iv) To study soil pollution-related aspects and to advise preventive as well as remedial measures for a safe food-chain.

3.3 Collection and Processing of a Composite Soil Sample

Objective

a) Collection of a composite soil sample

b) To air dry and process the composite soil samples collected for analysis in the laboratory

Soil is a heterogeneous body which is more in the vertical dimension than in the horizontal so, the entire field is thoroughly surveyed before collection of soil sample and is devided into sampling units on the basis of the variations in the following characteristics.

1. Physiography 3. Structure

2. Texture 4. Colour 5. Previous cropping history

Maximum area under one sampling unit should not exceed 2 hectare or 5 acre and one composite sample is collected for each sampling unit. Samples are collected randomly from different spots well distributed throughout the field. For this purpose 'Z' shaped imaginary lines are drawn and samples are collected from different spots located on these lines (Fig. 3.1). For one acre of area samples are to be collected from minimum 16 to 20 spots.

Soil samples are usually collected from 0 to 15 cm depth since the roots of most of the field crops derive nutrients and moisture from this depth. For Fruit crops sampling is to be done as per the type of root or the from the rhizophere zone.

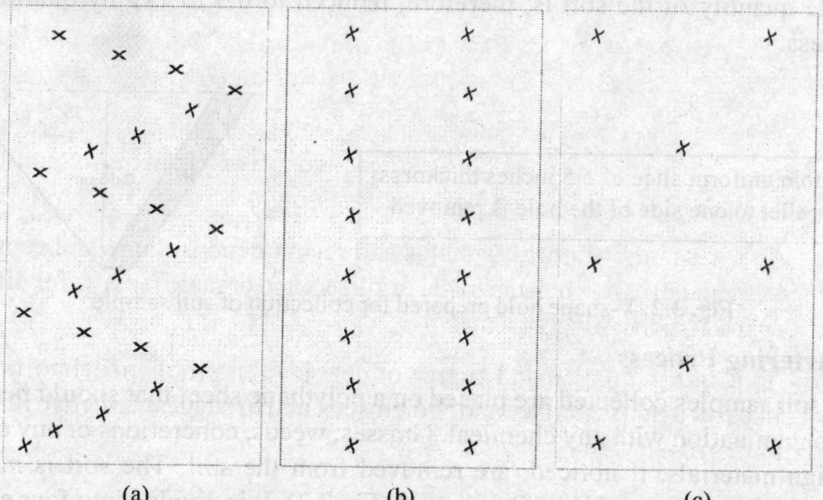

Fig. 3.1: (a)'Z' Shaped imaginary lines are drawn. Spots are selected on the lines to draw soil samples. (b) and (c) are wrong methods of selection of spots to draw soil samples.

Implements required

1. Spade
2. Khurpi
3. Screw auger
4. Soil sampling tube
5. Bucket
6. Tray / polythene sheet
7. Polythene bag
8. Drawing sheet
9. Pencil
10. Thread

If the soil is dry and hard, samples are collected by a spade. Before sampling, remove the grasses, weeds, concretions from the spot selected to draw the sample. While removing the grasses, weeds etc. care should be taken for least disturbance of the soil surface. A 'V' shape hole to a depth of 15 cm is then prepared by a spade (Fig.3.2) and a thin uniform slice of 1.5" thickness parallel to one side of the hole is removed and collected in a bucket or tray. Total quantity of the soil becomes large after drawing samples from a number of spots, whereas only 0.5 to 1 kg of soil is required for laboratory analysis. The whole quantity of the soil is, therefore, reduced to 0.5 to 1kg by quartering process.

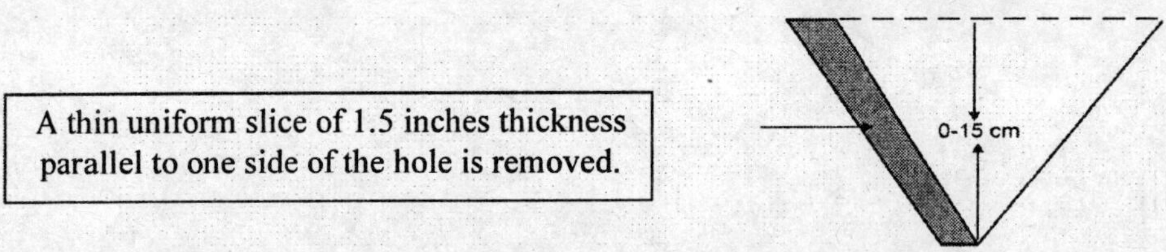

Fig. 3.2: V-shape hole prepared for collection of soil sample

Quartering Process

The soil samples collected are placed on a polythene sheet that should be free of contamination with any chemical. Grasses, weeds, concretions or any other foreign materials, if noticed, are removed from the soil. The soil is mixed thoroughly and a round heap is prepared (Fig.3.3). It is divided into four equal quarters and diagonally two opposite quarters are removed. The other two opposite quarters are mixed. This process is repeated till about 0.5 to 1 kg of soil is left. This is the composite soil sample which is kept in a polythene or cotton bag.

A piece of drawing sheet containing the following information written by pencil is inserted into the bag.

1. Name of the cultivator
2. Location
   a) Village / Town
   b) Block
   c) District
3. Type of land
   a) Low land/Medium land /Upland
   b) Virgin or cultivated land
4. Depth of collection
5. Date of collection
6. Previous crops grown and management practices followed if any
7. Signature of the collector

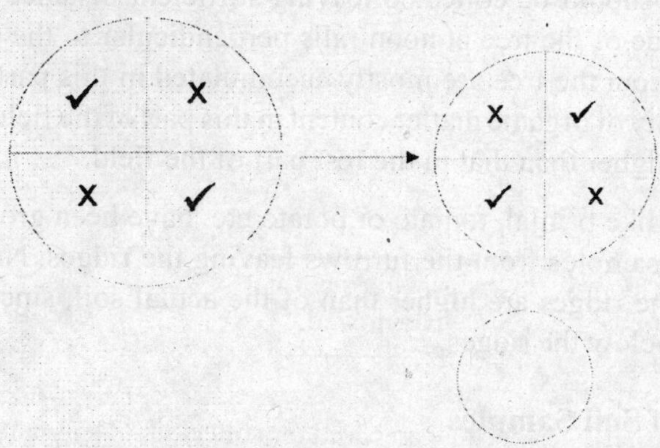

× = Portion of the soil to be removed. The two cross marks are located diagonally.

√ = Portion of the soil to be retained

Quantity of the soil is reduced by half after each quartering.

Fig. 3.3: Quartering process for reducing quantity of the soil sample

Screw auger is used to collect the soil sample when the soil is moist. The auger is inserted into the desired depth (0-15 cm) by rotating it clockwise and then pulled out of the soil. The soil adhered to the spiral end of the auger is removed and collected from a large number of spots in short time with less labour. Because a small quantity of soil is collected from each spot by the screw auger, quartering process may not be needed. Put a piece of drawing sheet written with the information as mentioned above into the bag containing the collected soil sample.

If the soil is wet, soil sampling tube is used to collect the soil sample. It is pressed into the desired depth of the soil and then pulled out of the soil. The soil entering inside the tube is removed and collected.

The bag containing the composite soil sample is to be tied by a thread or rubber band and should be immediately send to the laboratory for necessary analysis.

Precautions

1. Don't collect the soil sample near the bonds. Collect samples leaving sufficient distance from the bonds.
2. Don't collect the soil sample from or near a manure pit, heap or swampy area in the field. Samples are to be collected leaving sufficient distance from these spots.
3. Samples should not be collected from the area under shade of the tree at noon. It should be collected leaving sufficient distance from this area. The shade of the tree at noon falls perpendicular to the ground. Leaves falling from the tree are mostly accumulated in this part. Therefore the possibility of organic matter content in this part of the field and the nearby area is higher than that in the rest part of the field.
4. If crops like brinjal, tomato or potato etc. have been grown then collect the soil samples from the furrows leaving the ridges. Nutrient contents below the ridges are higher than of the actual soil, since fertilizers are placed below the ridges.

Processing of Soil Samples

Implements required

1. Polythene sheet/ paper sheet
2. Wooden hammer
3. 2 mm sieve
4. Polythene bottle of 500 g to 1 kg capacity

Air-drying of the soil

Spread the soil sample thinly on a polythene or paper sheet. Remove the grasses, weeds, concretions or any other foreign materials if noticed. Dry the soil under shade till its moisture content is in equilibrium with the moisture content of the atmosphere. This is called air-drying of soil. It takes nearly 5 to 7 days for air-drying.

Grinding of the soil

The soil sample is ground by a wooden hammer and is passed through a 2 mm sieve. The soil particles passing through the sieve are collected and preserved in the polythene bottle for laboratory analysis. The soil particles which pass through the 2 mm sieve are having the diameter of 2 mm or less and are called the fine earth fraction of soil. The particles retained on the sieve are having the diameter of > 2 mm and are called the coarse fragments. The fine earth fraction of the soil is collected leaving the coarse fragments.

Precautions

1. The soil is to be dried under shade but not under sun because of the possibility of the loss of organic matter which may bring a change in the physical and chemical properties of the soil.
2. For grinding the soil, use of iron hammer is avoided because of the possibility of contamination of soil with iron, which interferes in various chemical analysis like organic carbon content, available Fe content in soil etc.

N.B. Thickness of the metallic wires encircling the openings or meshes decreases proportionately with increase in the number of meshes per linear inch so that the value 0.37″ occupied by the wires encircling the meshes in one linear inch remains almost constant.

3.4 Soil Texture

Soil texture is the relative proportion of sand, silt and clay in soil by weight. As per the classification of the inorganic soil particle by the International Society of Soil Science, diameters of the sand, silt and clay particles vary from 2 to 0.02 mm, 0.02 to 0.002 mm and less than 0.002 mm, respectively. There are 3 broad textural groups i.e sand, loam and clay and each broad textural group is subdivided into different textural classes depending upon the proportion of sand, silt and clay are given below.

Broad textural group	Basic soil textural class	Texture
Sand or sandy soil (contains ≥70% sand and ≤ 15% clay by weight	SandLoamy sand	Coarse textured soil
Loam or loamy soil	Sandy loam	Moderately coarse textured soil
	Loam, Silt loam, Silt	Medium textured soil
	Sandy clay loam, Silty clay loam, Clay loam	Moderately fine textured soil
Clay or clayey soil (contains ≥ 35% clay by weight)	Sandy clay Silty clay Clay	Fine textured soil

Soil texture is the first identifying character of soil. It provides many useful information on the soil properties like its water holding capacity, nutrient retention capacity, stickiness, plasticity etc. Soil texture indicates the suitability of the soil for growing a crop. Although the accurate textural class of the soil can be determined by Bouyoucos hydrometer method but it is time consuming and may not serve the purpose of the farmer who wants immediate information on his soil before growing a crop. Feel method, in this case, is extremely useful to determine the soil texture immediately on the spot and to provide information on his soil before growing the crop. This method is more an arts rather than science since the accuracy of the determination of the textural class by feel method depends on the experience of the person analysing. The students are, therefore, advised to practice a number of samples to make themselves well acquainted with this method.

3.4.1 Determination of soil texture by Feel method

Procedure

1. Take some crushed soil (30-40 g) in a watch glass.
2. Make a round heap. Add water slowly to make the soil moist.
3. Then knead the soil thoroughly by the fingers.
4. All the characteristics like feel by finger, ball formation, roll formation, ring formation and ribbon formation and stickiness are to be tested with this kneaded moist soil.

Feel by finger

Rub the soil between the thumb and fore finger. Observe the feeling of the soil. The sand particles are gritty to feel. Silt particles are smooth or buttery to feel and clay particles are sticky to feel. If the soil is entirely gritty to feel, the texture may be sand or loamy sand. If the soil is sticky to feel it may be clay loam or clay. If the soil is gritty as well as sticky to feel it may be sandy clay loam or sandy clay. If it is very smooth and sticky it may be silty clay loam or silty clay. Sand and loamy sand can be distinguished from each other by ball formation. Similarly clay loam and clay ; sandy clay loam and sandy clay; silty clay loam and silty clay can be distinguished from each other by roll, ring and ribbon formations as well as stickiness (Table 3.1).

Ball formation

Form a ball with the kneaded moist soil. Taste the strength of the ball by pressing between the forefinger and thumb and observing the pressure needed to break the ball.

Roll formation

Roll the ball between the palms of the hand and observe the length of the roll formed.

Coarse textured soil : Roll formation does not occur

Moderately coarse textured soil : Formation of short roll occurs

For all medium textured, moderately fine textured and fine textured soils : Formation of long roll occurs

Ring formation

Try to form ring by bending the roll.

Coarse textured soils : Ring formation does not occur

Moderately coarse textured soils : Ring formation does not occur

Medium textured soils : Little tendency for ring formation

Moderately fine textured soils : Ring formation occurs but there are cracks at the outer surface of the ring

Fine textured soils : Ring formation occurs with smooth surface

Ribbon formation

Squeeze the ball between the fore finger and thumb to make a ribbon as long as possible until it breaks by its own weight.

Coarse textured soils : No ribbon forms

Moderately coarse textured and medium textured soils : Ribbon forms but the length of the ribbon is < 2.5 cm

Moderately fine textured soils : Ribbon forms with length 2.5 to < 5 cm

Fine textured soils : Ribbon forms with length ≥ 5 cm

Stickiness

To determine the stickiness little more water is added to the moist soil to make it slightly wet. Press the wet soil between the thumb and the fore finger. Throw the hand so that the soil mass outside the finger is removed. Now separate the fingers and observe the stickiness of the soil.

Non sticky : Soil mass does not adhere to fore finger or thumb. If adheres, it comes out (i.e. removed from the fingers) cleanly by slightly throwing the fingers. The coarse textured soils are non sticky.

Slightly sticky : Soil mass adheres to the fingers but comes out by moderately throwing the fingers. The soil shows little stretching while separating both the fingers. Moderately coarse textured and medium textured soils are slightly sticky.

Sticky : Soil mass adheres to both fore finger and thumb and tends to stretch when both the fingers are separated. Soil mass does not come out by throwing of the fingers. Moderately fine textured soils are sticky.

Very sticky : Soil mass adheres strongly to both fore finger and thumb and is decidedly stretched when both the fingers are separated. Soil mass does not come out even with violent throwing of the fingers. Fine textured soils are very sticky.

3.4.2 Soil texture by Bouyoucos hydrometer method

Principles

Determination of soil texture mainly depends on the complete dispersion of soil aggregates into their individual primary particles, accurate fractionation of the sample into various separates, sedimentation of different particles in the viscous medium as well as measurement of individual particle by the hydrometer. Chemical treatment of the soil is necessary to overcome the effect of the cementing agents such as organic matter, soluble salt, free iron as well as aluminium oxides and flocculating ions such as calcium and magnesium. The most common reagent for removing organic matter from soil is hydrogen peroxide. However, sodium hypochlorite solution at pH 12 was more effective in destroying organic matter. For removing calcium carbonate which is a cementing agent, the sample is treated with 1 N sodium acetate buffered at pH 5 followed by repeated washings.

Complete dispersion of soil particles is achieved by combination of mechanical agitation by the help of mechanical stirrer and chemical such as sodium hexametaphosphate or sodium oxalate (calgon). After complete dispersion, particle size is determined by the help of hydrometer as per the sedimentation principle of Stokes' law.

Table 3.1 Properties of soil textural classes

Textural class	Fill by finger	Ball formation	Roll formation	Ring formation	Ribbon formation	Stickiness
Sand	Very gritty	No	No	No	No	Nil
Loamy sand	- do -	Ball forms, but breaks with little pressure	No	No	No	Nil
Sandy loam	Gritty	Ball forms, but resists more pressure than loamy sand	Short rolls	No	<2.5 cm	Slightly sticky
Loam	Neither very gritty nor smooth	Forms firm ball*	Long rolls	Little tendency for ring formation	- do -	- do -
Silt loam	Smooth	- do -	- do -	- do -	- do -	- do -
Sandy clay loam	Sticky & gritty	- do -	- do -	Ring forms with cracks at the outer edge	2.5 < 5 cm	Sticky
Silty clay loam	Sticky & smooth	- do -	- do -	- do -	- do -	- do -
Clay loam	Sticky	- do -	- do -	- do -	- do -	- do -
Sandy clay	Gritty & sticky	Forms strong ball**	- do -	Ring forms with smooth surface	≥5 cm	Very sticky
Silty clay	Smooth & sticky	- do -	- do -	- do -	- do -	- do -
Clay	Sticky	- do -	- do -	- do -	- do -	- do -

* Breaks with moderate pressure; **Breaks with much pressure

Stokes' law states that

$$V = \frac{2g(\rho_p - \rho_w)r^2}{9 \quad \eta}$$

Where

V = Terminal velocity of settling of the particle (cm sec^{-1})

g = Acceleration due to gravity (cm sec^{-2}) (981 cm sec^{-2})

ρ_p = Particle density of soil (Mg m^{-3})

ρ_w = Density of fluid (Mg m^{-3})

r = Radius of the particle (cm)

η = Coefficient of viscosity of the fluid (Poise)

Equipments/Apparatus required

1. Bouyoucos hydrometer
2. Measuring cylinder (1000 cc)
3. Beaker (500 cc)
4. Glass rod
5. Balance
6. Electrical stirrer
7. Thermometer
8. 5% sodium hexametaphosphate or 0.5 N Na-oxalate

Procedure

Take 40 g of soil in a clean 500 mL polythene beaker. Add add 200 mL of distilled water followed by 15 mL of 5% sodium hexametaphosphate or 15 mL of 0.5 N sodium (Na) oxalate solution. Stir the contents for 5 minutes by a glass rod and another 20 minutes by an electrical stirrer. Transfer the contents to a 1 litre measuring cylinder without any loss and make up the volume to 1 litre by adding distilled water. Shake the contents end to end for ten times by closing the mouth of the cylinder by a rubber pad or palm of the hand. Place the cylinder immediately on a table and note the initial time of sedimentation that is when the cylinder is exactly placed on the table. Take the 1st hydrometer reading just after 5 minutes of the initial time of sedimentation. Then measure the temperature of the soil suspension by inserting a thermometer. Again shake the contents in the cylinder end to end for ten times as before. Place the cylinder immediately on the table and note the initial time of sedimentation. Take the 2nd hydrometer reading as well as the temperature of the soil suspension just after 5 hours of the initial time of sedimentation.

Calculation

1st Hydrometer reading of soil suspension at 5 minute = h_1

1st Temperature reading of soil suspension at 5 minute = t_1 °C

2nd Hydrometer reading of soil suspension at 5 hour = h_2

Temperature reading of soil suspension at 5 hour = t_2 °C

The hydrometer reading directly gives the density of the suspended material in g/litre. The hydrometer is calibrated at 20°C and needs temperature correction for increase in temperature above 20°C or decrease in temperature below 20°C of the soil suspension. A temperature correction factor of 0.3 is to be added with the measured hydrometer reading for each one degree rise in temperature above 20°C and 0.3 is to be subtracted for each one degree decrease in temperature below 20°C of the soil suspension to obtain the corrected hydrometer reading.

1st corrected hydrometer reading $H_1 = h_1 + (t_1 - 20°C) \times 0.3$

2nd corrected hydrometer reading $H_2 = h_2 + (t_2 - 20°C) \times 0.3$

1st corrected hydrometer reading (H_1) indicates the amount of (silt + clay) in gram in 1 litre of suspension.

$$\% \text{ (silt + clay) present in the given soil sample} = \left(\frac{H_1}{40} \times 100\right) \text{g}$$

% sand in the given soil sample = 100 – % (silt + clay)

2nd corrected hydrometer reading (H_2) indicates the amount of clay in gram in 1 litre of suspension.

$$\% \text{ clay in the given soil sample} = \left(\frac{H_2}{40} \times 100\right) \text{g}$$

% silt in the given soil sample = % (silt + clay) – (% clay)

The textual class of the given soil sample is then determined with help of textural triangle after calculating % sand, silt and clay in soil.

Precaution

1. The cylinder once placed on the table must not be disturbed unless the hydrometer reading is taken.
2. The hydrometer should be inserted into the soil suspension before 30 seconds of the pre determined time.
3. If frothing occurs and makes it difficult to note the hydrometer reading, add 2-3 drops of amyl or capryl alcohol to prevent frothing.

3.4.3 Soil texture by Robinson's International pipette method

Principle: Same as described in hydrometer method of soil texture analysis.

Apparatus required

500 ml beaker, hot water bath, rubber tipped glass rod, wash bottle, filter paper, electric stirrer, rubber pestle, 70 mesh sieve, measuring cylinder (1L capacity), glass funnel, thermometer, pipette (10 mL capacity, preferably Robinson's pipette), porcelain dishes or silica crucibles, balance and oven.

Reagents

1. Hydrogen peroxide (6%),
2. HCl (2%),
3. $AgNO_3$ (N/10),
4. NaOH (N/10) and
5. Phenolphthalein indicator

Procedure

Treatment with Hydrogen peroxide

1. Take 20g of air dry soil in a 500 mL beaker.
2. Add 50-60 mL of 6 % H_2O_2 and swirl the contents well. Let the reaction to proceed for 5-10minutes.
3. Place the beaker on a hot water bath.
4. Continue digestion, stirring the contents continuously with a rod to avoid frothing over, till the reaction completely subsides. In case frothing persists, add a second lot of 30-40 mL of H_2O_2 and see for effervescence, if any. Continue the step till the whole of the organic matter is oxidised.
5. Cool the beaker and rinse its sides with a rubber pestle and a jet of distilled water.

Treatment with acid and filtration

1. Add 25 mL of 2N HCl and shake to destroy $CaCO_3$ if present in the soil. If the soil contains more than 2% of $CaCO_3$, more HCl should be added at the rate approximately 25 mL for each percent of $CaCO_3$.
2. Dilute to approximately 250 mL and thoroughly rub the soil with a rubber pestle. Allow the content to react for about an hour with intermittent shaking.
3. Filter the contents through a buchner funnel under vacuum using What man No.50 filter paper.
4. Wash the soil with distilled water till the filtrate runs free of chloride (this can be tested with $AgNO_3$).

Dispersion and separation of coarse sand

1. Transfer the soil sample from the filter paper to a 500 mL beaker and make the volume to about 400-500 mL with distilled water. Add a few drops of phenolphthalein indicator.
2. Add NaOH till the whole suspension shows a pink colour. Indicating that its alkaline reaction (about 7 mL of N/10 NaOH is just sufficient for the purpose).
3. Stir the contents of the beaker with an electric stirrer for 10 minutes.
4. Transfer the content quantitatively to a 70 mesh sieve (0.2 mm), placed in the mouth of the funnel, which is held over the top of a 1L sedimentation cylinder.
5. Pour the suspension onto the sieve and with a jet of distilled water, wash as much as possible through the sieve until no more clay and silt remain on the sieve and cylinder is about one half full.
6. Transfer the coarser particles retained on the sieve to a weighed dish or crucible and dry it at 105°Cin the oven to a constant weight. Record the weight nearest to 0.001 g accurately.

Separation of silt and clay

1. Make up the volume of the suspension in the cylinder to 1L by adding more distilled water and homogenise the suspension.
2. Next day in the early morning hour, place the cylinder in a temperature controlled chamber to ensure minimum variation of temperature between the two sampling.

3. A little before the time of sampling of the suspension, stir the content with the plunger by moving it up and down, about 20-25 times in one minutes, so as to ensure stability of the suspension.

4. Remove the plunger gently and note down the time at which the plunger is taken out. This is the time of commencement of sedimentation.

5. Insert the sampling pipette gently into the suspension and dip it to 10cm depth from the surface of the suspension, about 20 seconds before the time is up and withdraw 10 ml suspension carefully.

6. Deliver the content of the pipette into a weighted dish and dry the sample at 105°C to a constant weight. Record the weight nearest to 0.001 g accurately.

Separation of clay

1. Stir the content of the cylinder again for one minute as before.

2. Sample in the same way at a depth of 10 cm immediately after the expiry of the requisite time for clay.

3. Transfer the suspension to a pre-weighed dish, dry the sample at 105°C to a constant weight. Record the weight nearest to 0.001 g accurately.

Separation of fine sand

1. Decant the bulk of the suspension and transfer the sediment, containing the fine sand, to a 500 mL beaker with a jet of distilled water.

2. Pour away the turbid suspension, each a time the sediment is washed with distilled water.

3. Continue the last step until the liquid no longer stands turbid.

4. Remove quantitatively, with a jet of distilled water, the particles, left in the beaker, to a weighed dish, dry the sample at 105°C to a constant weight. Record the weight nearest to 0.001 g accurately.

5. Compare the fine sand, thus, obtained with that obtained by the difference method.

6. Take a given amount of the same soil and record its oven dry weight for determination of water content of the soil.

Observations and Calculations:

a. Wt. of soil taken : 20 g

b. Water content of the soil : ------ %

Thus oven dry weight of the soil: $\frac{100 \times 20}{100 + b}$ = W g (say)

c. Wt. of dish : W_1 g

d. Wt of dish + coarse sand = W_2 g

Thus % coarse sand, $P = \frac{W_2 - W_1}{W} \times 100$

e. Temperature of the suspension: T^oC

f. Time at which the plunger is taken out the cylinder : t_o

g. Time at which the sample for silt + clay is to be collected : $t_o + t_{sic}$ (hr.min.sc.)

h. Time at which the sample for clay is to be collected: $t_o + t_c$ (hr.min.sc.)

At $t_o + t_{sic}$

i. Wt of dish: W_3 g

j. Volume of suspension taken for analysis: 10 mL

k. Wt of dish and silt + clay : W_4 g

l. Wt of silt + clay : $(W_4 - W_3)$ g

% silt + clay, $P_{sic} = \left(\frac{W4 - W3}{W} \times \frac{1000}{10} \times 100\right)$

At $t_o + t_c$

m. Wt of dish: W_5 g

n. Volume of the suspension taken; 10 mL

o. Wt of clay dish: W_6 g

p. Wt of clay alone: $(W_6 - W_5)$ g

Thus % clay $P_c = \left(\frac{W6 - W5}{W} \times \frac{1000}{10} \times 100\right)$

Therefore % silt , $P_{si} = P_{sic} - P_c$

q. Wt of dish : W_7 g

r. Wt of dish + fine sand : W_8 g

% of fine sand, P_{fs}

1. Experimental $= \left(\frac{W8-W7}{W}\times 100\right)$
2. By subtraction $= 100 - (P_{si} + P_c)$

3.5 Bulk Density, Particle Density and Pore Space in Soil

Bulk density

It is the oven – dried weight per unit volume of soil including pore space. It is also called as the apparent density of soil.

Particle density

It is the oven – dried weight for unit volume of soil excluding the pore space or it is the weight per unit volume of soil solid. It is also called as the true density of soil.

Both bulk density and particle density are expressed in g cm^{-3} or Mg m^{-3}. Generally bulk density of soils lies between 1.1 and 1.8 Mg m^{-3} and particle density ranges between 2.6 and 2.7 Mg m^{-3}. Bulk density of a soil is always less than their particle densities as in the former case pore space is included.

Pore space or porosity

Porosity is an index of the relative pore volume in the soil or volume fraction of pores. Its value generally lies in the range of 0.3-0.6 (30-60%). Coarse textured soils are less porus than fine textured soils, though the mean size of individual pores is greater in the former than in the latter. The total porosity, in any case, reveals nothing about the pore size distribution which itself is an important property.

3.5.1 Determination of bulk density (B.D.) in soil by clod method

Principle

Bulk density or apparent density of soil is the mass of oven dry soil per unit volume of soil including pore space. Its value is expressed in Mg/m^3 and may lie between 1.2 to 1.8 Mg/m^3 for mineral soils. Bulk density is determined by following two methods: (a) clod method, (b) core method. If a core sample of known volume is used, the determination of bulk density is simply a matter of drying the contents of the sampler and weighing.

In case of a soil clod of irregular shape, the bulk density is determined by dividing its weight in air with the loss in weight of the clod in water.

Apparatus

a) Physical balance

b) Moisture box

c) Hot air oven

d) Desiccator

e) Container for melting paraffin wax

f) Brush

g) Thread

Reagent : Paraffin wax (Sp. gravity 0.9 ± 0.01)

Procedure

1. Break the clod into two pieces. Determine the moisture content of one clod by gravimetric method after drying in hot air oven at 105°C for 24 hours.
2. Take another piece of soil clod of approximately 20 g. Brush off any loose materials adhering to it. Tie it with a strong thread. Weigh the clod along with the thread.
3. Dip the clod quickly for several times in the melted paraffin at 65°C to ensure a complete coating without any trapped air bubbles.
4. Weigh the clod coated with paraffin wax by hanging it in air. Take the weight of the clod once again while immersed in water.

Calculation

Weight of the air dry soil clod = x g.

Weight of the oven dry clod = y g.

Water content, M (%) = $\left(\frac{x-y}{y}\right)\times 100$

Weight of the air dry soil clod, without paraffin = A g.

Weight of air dry soil with paraffin wax coating in air = B g.

Weight of clod with paraffin wax coating in water = C g.

Weight of paraffin wax = (B – A) g

Let the density of paraffin wax be = 0.9 mg/m3

So volume (Vp) of paraffin wax = $\left(\frac{B-A}{0.9}\right)$ cm3

The apparent loss in weight of paraffin coated soil clod = (B - C) g

Let the density of water be 1 g/cm3

Volume of soil clod and paraffin wax = $\left(\frac{B-C}{1.0}\right)$ cm3

Volume (Vs) of soil clod (including pore space) = $\left(\frac{B-C}{1.0}-Vp\right)$ cm3

Weight of soil clod on oven dry basis = $\left(\frac{100A}{100+M}-Vp\right)$ g

The bulk density (Mg/m3) = $\left(\frac{100A}{100+M}\right)$/ Vs

3.5.2 Determination of bulk density (B.D.) in soil by core method

Principle

The bulk density expresses the ratio of a given mass of an oven dried soil to that of its volume including pore spaces. Bulk density of soils is closely related to total porosity. Generally, loose, porous and organic matter rich soil have lower bulk density. Sandy soils have relatively high bulk density(1.50 to 1.70 g/cm^3.) compare to slit and clay soils (1.25 to1.50 g/cm^3.) because total pore space in sandy soil is less than that of slit and clay soils. Bulk density typically increases with soil depth since subsurface layers have reduced organic matter, aggregation and root penetration compare to surface layers.

Materials required

1. Core sampler
2. Hammer
3. Khurpi
4. Knife
5. Electronic balance
6. Hot air oven
7. Scale
8. Slide caliper etc.

Procedure

1. Determine volume of the core available in core sampler with the help of slide calliper.
2. Select a representative spot in the field. Clean the surface vegetation and level it up.
3. Place the core inside the core sampler and put the core ring on it. Insert the core to the desired depth by hammering on the core ring.
4. Remove the soil from outside the core and lift the core from the base by the help of khurpi.
5. From both the ends of the core, trim off the excess soil carefully by a knife.
6. Put a filter paper at the base of the core. Weigh of the core along with the moist sample and put it in hot air oven at 105°C till to get a constant weight.
7. Then weigh of the dried sample and core.

Calculation

a. Volume of the core sampler = $V=\pi r^2h$(cc) =Volume of soil soil sample = Vs

b. Core sampler is cylindrical, so volume of core sampler is πr^2h

Where,

r = Inner radius of core sampler

h = Height of core sampler

W_1 = Weight of empty moisture box (g)

W_2 = Weight of moisture box along with moist soil sample(g)

W_3 = Weight of moisture box along with oven dried soil sample (g)

Bulk density = $\frac{W_3 - W_1}{V_s}$ (mg/m^3)

Sample problem: Calculate the bulk density from the following data. Fresh weight of soil = 2501g; Weight of water = 750g; Height of core = 10cm and Diameter of the core = 12cm

Observation table:

Depth of soil (cm)	Moisture box no.	Weight of empty moisture box W_1(g)	Weight of moisture box + moist soil sample W_2(g)	Weight of moisture box + oven dried soil sample W_3(g)	Radius of the core (cm)	Height of the core (cm)	Volume of the core (V) in cc	Bulk density $= \frac{W_3 - W_1}{V_s}$ (g/cc)
0-15								
15-30								
30-45								

Ans:

Volume of the core sampler =Volume of soil soil sample =Vs = $\pi r^2 h$

$$= 3.14 \times 6^2 \times 10 = 1130.4 \text{ cm}^3$$

$$\text{Bulk density} = \frac{W_3 - W_1}{V_s} = \frac{2501 - 750}{1130.4} = 1.549 \text{ g/cm}^3$$

3.5.3 Determination of particle density (PD) in soil by Pycnometer method

Principle

Particle density of the soil is otherwise known as specific gravity of soil. The particle density of a soil is the ratio of the mass to the volume of the soil particle excluding pore spaces. Particle density can be determined with relative density bottle or pycnometer bottle.

Apparatus

1. Moisture box
2. Pycnometer bottle
3. Water bath or hot plate
4. Balance
5. Hot air oven.

Procedure

1. Take approximately 20 g of soil in the moisture box and determine the water content of the sample gravimetrically.
2. Weigh a dry and clean empty pycnometer bottle. Put approximately 10 g of soil to the pycnometer bottle and weigh it.
3. Fill the pycnometer bottle with freshly boiled and cooled distilled water and heat the pycnometer bottle on a hot water bath with intermittent shaking.
4. After cooling the content, make up the volume with freshly boiled and cooled distilled water to remove air bubbles.
5. Wipe the excess water present on the outside wall of the bottle and weigh it.

6. Clean the bottle and fill the bottle with freshly boiled and cooled distilled water. Wipe excess water from the outside and take its weight.

Calculation

Weight of pycnometer bottle = A g

Weight of pycnometer bottle + Soil = B g

Weight of soil = (B –A) g

Let the water content in soil = W %

So weight of soil on oven dry basis $= \frac{100}{(100+W)}$(B – A) g = Ms

Weight of pycnometer bottle + weight of soil + weight of water present in pycnometer bottle = C g

Weight of pycnometer bottle + weight of water required to completely fill the pycnometer bottle = D g

Volume of soil Vs $= \frac{(D+Ms)-C}{Pw}$

Where Pw = Density of water

Particle density (Mg/m^3) Pd $= \frac{Ms}{Vs}$

3.5.4 Determination of Bulk density, Particle density and pore space in soil by cylinder method

Equipments Required

1. Physical balance
2. Measuring cylinder
3. Glass rod
4. Burette

Procedure

Weigh 40 g of soil and transfer to a 100 cc clean and dry measuring cylinder. Strike the cylinder containing soil ten times on the table from a height of 10 cm for uniform packing of the soil. After striking, note the volume of soil. Transfer the soil to a piece of paper carefully. Add water from the burette sufficient to immerse the soil in the cylinder. Carefully transfer the total amount of soil to the cylinder from the paper and stir with glass rod. Wash the glass rod inside the cylinder with water from the burette and remove the glass rod. Calculate the bulk density (BD), particle density (PD) and percent pore space of the soil as follows.

Calculation

a. Wt. of soil = w g

b. Volume of soil = V_1 cc

c. Volume of water added from burette = V_2 cc

d. Volume of soil water suspension = V_3 cc

e. Volume of soil solid = $(V_3 - V_2)$ cc

$$\text{Bulk density} = \frac{w\,g}{V_1 cc}$$

$$\text{Particle density} = \frac{w\,g}{(V_3 - V_2)cc}$$

$$\text{Percent pore space} = (1 - \frac{BD}{PD}) \times 100$$

Precaution

(1) A clean dry cylinder should be used

(2) The volume of the soil water suspension should not exceed the maximum graduated volume of the cylinder.

3.6 Determination of Water Holding Capacity (WHC) in Soil

Principle

The maximum water holding capacity is the amount of water, expressed as percentage on oven dry basis held by a soil when it is completely saturated with water. At this moisture content, all the pore spaces are completely filled with water.

Apparatus required

1. Keen Raczkowski box with perforated bottom
2. Filter paper
3. Spatula
4. Water tray

5. Balance
6. Watch glass

Procedure

Sieve the air dried and processed soil sample with 2 mm sieve. Put the filter paper on the perforated bottom of the keen raczkowski box. Transfer the soil in small lots by means of a spatula so as to fill the box. Tap the box gently with each addition of soil during filling process for uniform packing. When the box is full, slice off the extra quantity with the help of the sapatula. Again tap the box, add a little more quantity of soil and again remove the surplus soil to ensure a level surface.

Place the brass box containing soil in a water tray and leave it for saturation over night. Next day remove the box, wipe it dry from outside by the help of filter paper and weigh by physical balance. Place the box in the hot air oven at 105°C and dry it till to get a constant weight. Cool the brass box in a desiccator and weigh by the help of physical balance.

Apply correction for the amount of water absorbed by the filter paper as follows. Weight 10 filter papers together. Saturate them with water. Roll gently a glass rod over them to squeeze out water uniformly. Weigh again, calculate the average amount of water held by one piece of filter paper.

Calculation

a. Weight of brass box along with a filter paper placed inside the perforated bottom = a g.

b. Weight of brass box along with filter paper with saturated soil = b g.

c. Weight of brass box along with filter paper with oven dry soil = c g.

d. Amount of water held by one paper = d g.

$$\text{Maximum water holding capacity} = \frac{(b-c-d)\ g}{(c-a)} \times 100$$

3.7 Determination of Soil Moisture

Determination of soil water or soil moisture content of the given soil sample by gravimetric method is described in this section.

Principle

The soil water plays an important role in the physical, chemical and biological activities take place in the soil which directly or indirectly affects the crop growth. The water content of the soil can be measured by different methods. These methods are gravimetric method, neutron scattering method and electrical resistance method. In gravimetric method soil samples are collected and dried in the hot air oven at 105°C at least for 24 hours or till we get a constant weight. Water content can be expressed in mass basis (g/g or %) or volume basis (cm^3cm^{-3} or c.c. / c.c.).

Apparatus required

1. Soil auger
2. Aluminium moisture box
3. Hot air oven
4. Physical balance
5. Desiccator
6. Trenchoe

Procedure

Weigh a clean and dry soil moisture box made up of aluminium. Collect the moist soil sample by the help of soil auger from pre-determined depth of the study area. Keep the moist soil in the aluminium box and immediately close the lid of the box to avoid water loss by evaporation. Take the aluminium moisture box to the laboratory and determine its weight. Open the lid of the moisture box and dry continuously in a hot air oven at the temperature of 105°C at least for 24 hour. Allow the sample to cool inside the oven for some time. Close the lid of the box and place the sample in the desiccator for cooling. After cooling, weigh the moisture box along with the lid. Repeat this process for three times to get a concurrent reading. Determine the water content by mass basis as per the following calculation.

Calculation

Weight of the oven dry empty clean Aluminium moisture box with lid = A g

Weight of the moist soil including moisture box with lid = Bg.

Weight of the oven dry soil including moisture box with lid = Cg.

Water content in the soil = (B-C) g.

Weight of dry soil only = (C-A) g.

Water content (Mass basis) $= \left(\frac{B-C}{C-A}\right) g/g = M$

$$\text{Water content (\%)} = \left(\frac{B-C}{C-A}\right) \times 100 = M\%$$

If the bulk density of the given sample is known the volumetric water content (θ) can be calculated by the following formula.

$\theta = M \times \text{Bulk Density}$

3.8 Determination of Electrical Conductivity (EC) and pH in Soil

3.8.1 Electrical conductivity (EC) of Soil

Electrical conductivity of the soil is the quantity of electricity in coulombs that passes per second per cm^3 of the soil solution under an applied potential difference of one volt. It is written as EC in abbreviated form and is expressed in millimohs cm^{-1} or dS m^{-1} (deciSiemen m^{-1}). Current flows in the soil solution due to presence of soluble salts like Cl^-, SO_4^{2-} and NO_3^- of Na^+, K^+ Ca^{++} and Mg^{++}. These salts fully dissociate in water to their constituent cations and anions and the current is carried by those cations and anions in the soil solution. Therefore, electrical conductivity of soil increases with increase in soluble salt content in soil. A saline or saline- alkali soil can be identified from its EC value.

Equipments and Apparatus required

1. Conductivity meter
2. Physical balance
3. 100 mL plastic beaker
4. Glass rod
5. Thermometer
6. Filter paper

Reagent required

1. Standard 0.02 M KCl : Dissolve 1.491 g of analytical grade (AR) KCl (dried at 110°C in oven) in water and dilute to one litre. This solution has a conductivity of 2.768 mmhos cm^{-1} at 25°C.

Procedure

Take 25 g of soil in a 100 mL plastic beaker. Add 50 mL of distilled water (1:2 ratio). Stir the contents by a glass rod for several times during a period of 2 hours and then allow to stand till the soil particles settle down.

Decant or filter the supernatant to another 100 mL beaker. Dip the conductivity cell of the conductivity meter into the filtrate and measure the conductance. Note the temperature of the solution by a thermometer. Calculate the conductance of the soil suspension by multiplying with the temperature correction factor given below.

N.B.: In order to save time, conductance or EC of the soil solution can be measured following the pH measurement. Wait, after pH measurement, till the soil particles settle down and a clear supernatant is obtained above the settled soil particles. Dip the conductivity cell into the supernatant to measure the conductance.

Table 3.2: Temperature factor (ft) for determining conductivity at 25°C

EC at 25°C = EC at measured temp. × ft					
°C		Ft	°C		ft
15.0	-	1.247	31.0	-	0.890
16.0	-	1.218	32.0	-	0.873
17.0	-	1.189	33.0	-	0.858
18.0	-	1.163	34.0	-	0.843
19.0	-	1.136	35.0	-	0.829
20.0	-	1.112	36.0	-	0.815
21.0	-	1.087	37.0	-	0.801
22.0	-	1.064	38.0	-	0.788
23.0	-	1.043	39.0	-	0.775
24.0	-	1.020	40.0	-	0.763
25.0	-	1.000			
26.0	-	0.979			
27.0	-	0.960			
28.0	-	0.943			
29.0	-	0.925			
30.0	-	0.907			

3.8.2 Soil pH

The soil pH is the negative logarithm of the H^+ activity in the soil solution. It is the measure of soil acidity and alkalinity. Soil pH significantly controls nutrient availability in soil. The optimum soil pH is 6.0 to 6.5 for availability of most of the nutrients.

Requisites

1. pH meter
2. 25 mL Pipette
3. 50 mL polythene beaker

4. 100 mL wide mouth polythene bottle
5. Glass rod

Reagent required

Buffer solutions of pH 4 and 9.2: Dissolves pH buffer tablets of 4.0 and 9.2 separately in 100 mL distilled water to prepare the buffer solutions of pH 4.0 and 9.2. Buffer solution of pH 9.2 can be prepared by dissolving 0.381 g of sodium tetraborate ($Na_2B_4O_7.10H_2O$) in distilled water and diluting to 100 mL. Store the buffer solutions in wide mouth polythene bottles. Discard the buffer solution and prepare a fresh one if any mould formation is obtained.

Procedure

Take 10 g of soil in a 50 mL polythene beaker. Add 25 mL of distilled water by a 25 mL volumetric pipette. Stirre the contents by a glass rod for 1 to 2 minutes. Then wait for 30 minutes with intermittent shaking to attain equilibrium between the adsorbed H^+ and active H^+. Then measure the pH of the soil suspension after standardizing the pH meter with pH 4.0 and 9.2 buffer solutions. The soil suspension should be stirred just before measuring the pH. The electrodes should be washed properly just before and after measuring the pH.

Table 3.3 pH Range – Soil Reaction Rating

pH Range	Soil Reaction Rating
<4.6	Extremely acid
4.6-5.5	Strongly acid
5.6-6.5	Moderately acid
6.5-7.0	Neutral
7.1-8.5	Moderately alkaline
>8.5	Strongly alkaline

3.8.3 Colorimetric Estimation of Soil pH

Colorimetric estimation of soil pH is an approximate value of soil pH. In this method a colour is developed by adding a suitable reagent to soil suspension. By using universal pH indicator one can determine the soil pH. This method is useful for immediate soil assessment.

Reagents

Universal pH indicator (pH 4.0 to 11.0)

Apparatus required

1. Test tube
2. Test tube stand
3. 10 mL volumetric pipette

Procedure

Take a clean dry test tube. Add $BaSO_4$ and soil to a total height of 2 cm in the following proportion depending upon the texture of soil.

Soil texture	$BaSO_4$		Soil
sand, loamy sand	1	:	3
sandy loam, loam, silt loam, silt	1	:	1
sandy clay loam, silty clay loam, clay loam	2.5	:	1.5
sandy clay, silty clay and clay	3	:	1

It is better to take soil after taking $BaSO_4$ in the test tube. Add 10 mL of distilled water followed by 10 to 15 drops of universal pH indicator. Shake the contents end to end vigorously for 2 minutes to mix the contents thoroughly. Then keep the test tube in a test tube stand and wait till the soil particles settle down and a clear transparent solution is obtained above the settled soil particles. Match the colour of the solution with the colour chart provided on the indicator bottle and determine the pH of the solution. A clear supernatant is not obtained if the clay particles do not settle and remain dispersed. Under that case, add more of $BaSO_4$. This causes flocculation of the colloidal particles and a clear transparent solution can be obtained.

Why $BaSO_4$ is added

1. $BaSO_4$ acts as a coagulant. It causes flocculation of the colloidal particles to get a clear solution. If $BaSO_4$ is not added, the clay particles remain dispersed and a clear transparent solution is not obtained for measuring pH.
2. Since $BaSO_4$ is a spairingly soluble salt, therefore it dissociates to very less extent and does not replace more of the adsorbed H^+ to the soil solution to get a lower pH value than that should be under the actual field condition.

$CaCl_2$ also acts as a coagulant. Since it is a strong electrolyte it completely dissociates to Ca^{++} and Cl^-. The Ca^{++} replace more of the adsorbed H^+ to the soil solution to give a lower pH value than that should be under field condition.

3.9 Estimation of Redox Potential in Soil

The importance of pH as a "master variable" controlling chemical reactions in soils has been discussed in various equilibrium reactions. However, soils subjected to fluctuations in water content come under the influence of another master variable, the *reduction-oxidation (Redox) potential*. Under conditions of water saturation, the lack of molecular oxygen can result in a sequence of redox reactions that changes the soil pH. In this sense the redox state of the soil exerts control over the pH. The nature of redox reactions profoundly influence metal ion solubility and the chemical form of ions and molecules dissolved in soil solution.

For any electron transfer half reaction of general form is

Oxidixed molecule $+ mH^+ + n$ electrons = reduced molecule

The Nernst equation can be written as;

$$E_h = E_h^0 - \frac{0.059}{n} \log \frac{\text{(reduced molecule)}}{\text{oxidized milecule}) \times (H^+)^m}$$

Where;

E_h = Potential for the half reaction (volts, V)

E^o_h = staential state potential for the haf reaction, (volts, V)

n = moles of electrons involved in the reaction as written

m = moles of protons involved in the reaction as written

and the parenthesis denotes chemical activities of the reduced and oxidized molecules.

3.9.1 Redox Potential Measurement in Soil by Electrode

Theory

The commonest method of measuring *redox potential* is to immerse a platinum (Pt) electrode along with a reference electrode into a solution. The electrodes are then connected to a potentiometer that measures the potential difference between the two electrodes. Reduced species in solution tend to donate electrons to the conducting Pt electrode, while oxidized species tend to accept electrons from the electrode. This creates electron flow in the electrode. However, to cause any detectable movement of electrons in the electrode, the reaction between the *oxidized (ox)* and *reduced (red)* form of an electron in solution must be shifted away from equilibrium. At the steady state of equilibrium, the forward reaction rate equals the backward rate and the net electron flow (*net current*) is zero. Even at equilibrium, however, electron flow in either direction, termed as exchange current is not zero. A very small shift of the electrode potential away from its equilibrium value causes the half reaction to proceed either to the left or right, thereby creating a net current that can be measured. How precisely this measurement can determine the equilibrium potential of a particular half reaction depends on how steeply the net current deviates from zero near the equilibrium potential. The greater the magnitude of the exchange current, the more steep is the net current function. This, in turn is a function of the concentration of reduced and oxidized species near electrode surface.

Requisites

1. pH meter having Platinum (Pt) electrode
2. 25 mL Pipette
3. 50 mL polythene beaker
4. 100 mL wide mouth polythene bottle
5. Glass rod

Reagent required

Buffer solutions of reduced pH 4 and 9.2: Dissolves pH buffer tablets of 4.0 and 9.2 separately in 100 mL distilled water to prepare the buffer solutions of pH 4.0 and 9.2 and a fresh one is prepared if any mould formation is obtained. This is used to standardize the instrument.

Procedure

Take 10 g of reduced soil in a 50 mL polythene beaker. Add 25 mL of distilled water by a 25 mL volumetric pipette. Stirr the contents by a glass rod for 1 to 2 minutes. Then wait for 30 minutes with intermittent shaking to attain equilibrium between the oxidized and reduced species. Then measure the *redox potential* of the soil suspension after standardizing the Pt electrode. The soil suspension should be stirred just before measuring the potential. The electrodes should be washed properly just before and after measuring the redox potential. Express the results in mV.

3.10 Organic Carbon Content (O.C.) in Soil

Organic matter in soil can be determined by different methods of which the method developed by Walkley and Black, is followed by the soil testing laboratories for routine analysis. It is also known as chromic acid digestion method. The organic matter content of the soil is estimated from the amount of organic carbon present, as this element represents, on an average 52-58 % of the organic matter. In order to find out the humified organic material, the organic carbon content is multiplied by 1.724, which represents the ratio between humus and organic carbon (100/58) and is known as the "Von Bemmlen factor".

3.10.1 Determination of organic carbon content (O.C.) in soil by Walkley and Black method

Principles

A known weight of soil is treated with an excess volume of standard $K_2Cr_2O_7$ in presence of concentrated H_2SO_4.The soil is slowly digested, by the heat of dilution of H_2SO_4 and the organic carbon in the soil is thus oxidised to CO_2. The highest temperature attained due to the heat generated on the addition of H_2SO_4 is approximately 120°C, which is sufficient to oxidise the active forms of the soil organic C, but not the mere inert form of carbon, that may be present. The excess of $K_2Cr_2O_7$ not reduced by the organic matter, is titrated, against a standard solution of a reducing agent Fe $(NH_4)_2$ $(SO_4)_2$ .6 H_2O, using diphenyl amine or ferroin indicator.

Reactions

$$K_2Cr_2O_7 + 4H_2SO_4 \rightarrow K_2SO_4 + Cr_2(SO_4)_3 + 4H_2O + 3O \quad (1)$$

$$1.5C + 3O \rightarrow 1.5CO_2 \quad (2)$$

$$K_2Cr_2O_7 + 1.5C + 4H_2SO_4 \rightarrow K_2SO_4 + Cr_2(SO_4)_3 + 1.5CO_2 + 4H_2O \quad (3)$$

The unreacted $K_2Cr_2O_7$ is then titrated against ferrous ammonium sulphate.

$$K_2Cr_2O_7 + 6\ Fe\ (NH_4)_2\ (SO_4)_2\ .6H_2O + 7H_2SO_4 \rightarrow$$

$$\text{(unreacted)} \quad K_2SO_4 + Cr_2\ (SO_4)_3 + 3\ Fe_2(SO_4)_3 + 6\ (NH_4)_2SO_4 + 43H_2O$$

Action of Indicator

$$2C_6H_5NHC_6H_5 \xrightarrow{+O} 2(C_6H_5.NH.C_6H_4) \xrightarrow{+O} C_6H_5N\text{-}C_6H_4C_6H_4N\text{-}C_6H_5$$

Diphenyl amine — Diphenyl benzidine (Violet)

Apparatus required

1. Analytical balance
2. 500 mL conical flask
3. 10 mL volumetric pipette
4. 100 mL and 500 mL measuring cylinders
5. Burette

Reagents required

1. *Standard 1 N $K_2Cr_2O_7$*: Dissolve 49.04 g of $K_2Cr_2O_7$ in distilled water and dilute to 1 litre.

2. *Ferrous ammonium sulphate solution (0.5 N)*: Dissolve 196.1 g of $Fe(NH_4)_2(SO_4)_2.6H_2O$ in 800 mL of distilled water containing 20 mL conc. H_2SO_4 and dilute to 1 litre.
3. *Diphenylamine indicator solution* : Dissolve 0.5 g of the indicator in 20 mL water and 100 mL conc. H_2SO_4
4. Conc. H_3PO_4 (85%)
5. Solid NaF
6. *Ferroin indicator*: Dissolve 1.485 g of O-phenanthroline monohydrate and 0.695 g of ferrous sulphate heptahydrate in water and adjust the volume to 100 mL.

Procedure

1. Take 2 g of soil passed through 60 mesh sieve, in a 500mL conical flask. Only 0.5 or 1.0 gram of soil is to be taken if the soil contains high amount of organic matter.
2. Add 10 mL of 1N $K_2Cr_2O_7$ followed by 20 mL of concentrated H_2SO_4.
3. Mix the contents by gentle swirling for 1 minute and keep for 30 minutes for completion of the redox reaction.
4. Then add 200 mL of distilled water and 10 mL of concentrated H_3PO_4 or 0.2 g of solid NaF (approximately).
5. Mix the contents and add 20-30 drops of diphenylamine indicator.
6. Titrate against 0.5 N ferrous ammonium sulphate solution till the initial violet colour of the indicator changes to the parrot green colour end point. Run a blank side by side following all the above steps without taking the soil.
7. If ferroin indicator is added instead of diphenylamine then NaF or H_3PO_4 need not be added. Add 4 to 5 drops of ferroin indicator after adding 200 mL of distilled water.
8. Then titrate against 0.5 N ferrous ammonium sulphate solution till a maroon colour end point is obtained. In the middle of the titration, green colour appears indicating that the titration is nearer to the end point. Then add the ferrous ammonium sulphate solution from the burette slowly to avoid excess end point.

Observations and Calculations

1. Weight of the soil : W g
2. Volume of 0.5 N $Fe(NH_4)_2(SO_4)_2.6H_2O$ solution consumed in blank titration: B mL
3. Volume of 0.5 N $Fe(NH_4)_2(SO_4)_2.6H_2O$ solution consumed in sample titration: S mL
4. % of Organic carbon in soil : $10\left(1-\frac{S}{B}\right)\times 0.0039\times\frac{100}{58}$
5. % of O.M. : % Organic carbon $\times\frac{100}{58}$

3.11 Available Nitrogen (N) in Soil

Principle

The easily mineralizable nitrogen in soil is estimated using alkaline $KMnO_4$ (Subbiah and Asija, 1956), which oxidizes and hydrolyses the organic matter present in the soil (Alkaline Potassium Permanganate ($KMnO_4$) Method). The liberated ammonia is condensed and absorbed in boric acid, which is titrated against standard H_2SO_4. The method has been widely adopted to get a reliable index of nitrogen availability in soil due to its rapidity and reproducibility. The process of oxidative hydrolysis is, however a progressive one and thus, a uniform time and heating temperature should be allowed for best results. Use of glass beads checks bumping while liquid paraffin checks frothing during distillation.

Equipment/Apparatus required

1. Nitrogen distillation unit (Kjeldahl distillation unit),
2. Heater or burner
3. Conical flasks
4. Pipettes
5. Burette

Reactions

During Distillation process

$$2\,KMnO_4 + 5H_2O \xrightarrow{NaOH} 2KOH + 2Mn(OH)_4 + 3O \quad (1)$$

$$3R.CH.NH_2COOH + 3O\ 3R.CO.COOH + 3NH_3 \quad (2)$$

$$2\,KMnO_4 + 3R.CH.NH_2COOH + 5H_2O \xrightarrow{NaOH} 2KOH + 2\,Mn(OH)_4$$

$+3R.CO.COOH + 3NH_3$

$NH_3 + H_3BO_3\ NH_4\ H_2BO_3$

During Titration $\longrightarrow$

$$NH_4H_2BO_3 + \frac{1}{2}H_2SO_4 \rightarrow \frac{1}{2}(NH_4)_2\ SO_4 + H_3BO_3$$

Reagents required

1. *0.32% $KMnO_4$*: Dissolve 3.2 g of $KMnO_4$ in water and make the volume to one litre.
2. *2.5% NaOH*: Dissolve 25 g of sodium hydroxide pellets in water and make the volume to one litre.
3. *2% Boric acid*: Dissolve 20 g of boric acid powder in warm water by stirring and dilute to one litre.
4. *Mixed Indicator*: Dissolve 0.066 g of methyl red and 0.099 g of bromocresol green in 100 ml of ethyl alcohol. Add 20 ml of this mixed indicator to each litre of 2% boric acid solution. Adjust the pH to 4.5 with dilute NaOH.
5. *0.02N H_2SO_4*: Prepare approximately 0.1N H_2SO_4 by adding 2.8 mL of conc. H_2SO_4 to about 1 litre of distilled water. From this, prepare 0.02N H_2SO_4 by diluting a suitable volume (20 ml made to 100 ml) with distilled water. Standardize it against 0.1N NaOH solution.
6. *0.1N NaOH.* Dissolve 4 gm NaOH in 1000 ml distilled water.

Procedure

1. Take 20 g of soil sample in a 800 ml Kjeldahl flask.
2. Add 20 ml of distilled water. Wash down the soil, if any, adhering to the neck of the flask.
3. Add 100 ml of 0.32% of KMnO4 solution followed by 250 mL of water.
4. Add a few glass beads.
5. Add 2-3 ml of paraffin liquid, avoiding contact with upper part of the neck of the flask followed by 200-300 mL distilled water.
6. Measure 20 ml of 2% boric acid containing mixed indicator in a 250 ml conical flask and place it under the receiver tube. Dip the receiver tube in the boric acid.
7. Run tap water through the condenser.

8. Add 100 ml of 2.5% NaOH solution and immediately attach to the rubber stopper fitted in the alkali trap.
9. Switch the heaters on and start distillation. Absorption of NH_3 by boric acid is indicated by the change of the initial pink or red colour to blue colour. Continue distillation until about 100 to 150 mL of distillate is collected. The completeness of distillation is checked by exposing moistened red litmus paper at the tip of the condenser. If the paper remains red, it indicates completeness of distillation/else continues distillation.
10. First remove the conical flask containing distillate and then switch off the heater to avoid back suction.
11. Titrate the distillate against 0.02N H_2SO_4 taken in a burette until pink colour starts appearing.
12. Run a blank following all the above steps except taking the soil.
13. Carefully remove the Kjeldahl flask after cooling and drain the contents in the sink.
14. Collects the glass beads for further use.

Calculation

Volume of acid used to neutralize ammonia in the sample = A – B ml

Available N (Kg/ha) = (A–B) x N x 1568

= (A–B) x 31.36

Where,

A = Volume of 0.02N H_2SO_4 consumed in the sample titration (mL).

B = Volume of 0.02N H_2SO_4 used in blank titration (mL).

N = Normality of H_2SO_4 (0.02 N)

Interpretation of Results

Amount of Available N (Kg/ha)	Comments
<280	Low
280-560	Medium
>560	High

[(A-B) x N x 0.014 x 1/20 x 2.24 x 10^6 = (S-B) x N x 14 x 0.5 x 2.24 x 10^2 = (A-B) x N x 1568)]

3.12 Available Phosphorus (P) in Soil

In soil testing, the available P content is readily determined by extraction with suitable reagents according to specified soil solution ratio and time of shaking. In the filtered extract, P is estimated spectrophotometrically by adding ammonium molybdate and thereafter reducing the molybdenum phosphate complex for the concentration of P in the test solution. Two methods are generally adopted for extraction of available P in soil depending upon the soil reaction. a) Bray's method for acidic soil, and b) Olsen's method for neutral or alkaline soil.

3.12.1 Determination of available phosphorus in soil by Bray's method

Bray's method No. 1 (Bray and Kurtz, 1945) for acid soils (pH around 5.5 or less)

Bray's method of determination of available phosphorus (P) is adopted for acid soils. Available P in acid soil includes the nonspecifically held P, specifically held P forming mononuclear bridge complex, Fe-and Al-P as $Fe(OH)_2 H_2PO_4$ or $Al(OH)_2 H_2PO_4$ in weakly crystalline or amorphous form. It also includes Ca-P if exist in soil. The chemical composition of Bray's P-I extract is 0.03N NH_4F in 0.025N HCl.

Principles

This method is used to determine the available P in acidic and strongly acidic soils. It mostly extracts P bound to Al and Fe. The fluorine anion (F) has special property of complexing and ions in acid solution with consequent release of P on these trivalent ions in soil. The soluble phosphate forms heteropoly complex with molybdate ion, phosphorous being the central co-ordinating atom. It gives a faint yellow colour to their water solution, which on reduction with stannous chloride or ascorbic acid give a blue colour, which is proportional to the concentration of phosphate in solution and is measured at 660 nm wavelength using spectrophotometer.

Reaction

With the extractant and soil containing phosphorus

$$3NH_4F + 3HCl + Al(OH)_2H_2PO_4 \rightarrow (NH_4)_3AlF_6 + H_3PO_4 + 2\ H_2O + 3NH_4Cl$$

$$3NH_4F + 3HCl + Fe(OH)_2H_2PO_4 \rightarrow (NH_4)_3FeF_6 + H_3PO_4 + 2\ H_2O + 3NH_4Cl$$

In the filtered extract, P is estimated colorimetrically by adding molybdate solution and stannous chloride solution.

$(NH_4)_6Mo_7O_{24}.4H_2O + HCl \rightarrow 7\ H_2MoO_4 + 6\ NH_4Cl$

Ammonium Molybdate Molybdic acid

$H_3PO_4 + 12\ H_2MoO_4 \rightarrow H_3P(Mo_3O_{10})_4 + 12\ H_2O$

Phosphate Molybdic acid Phosphomolybdate yellow coloured heteropoly complex

Phosphomolybdate complex + $SnCl_2$ → Reduced Phosphomolybdate
(Yellow coloured) (Blue coloured)

Equipments/Apparatus required

Spectrophotometer, Mechanical shaker, Pipettes, Bearkers/flasks

Reagents required

1. *Bray extractant No 1* (0.03N NH_4F in 0.025N HCl):

   a. 1 N NH_4F: Dissolve 37 g of NH_4F in distilled water and dilute to 1 litre.

   b. 0.5 N HCl: Dilute 21.6 mL of conc. HCl to a volume of 500 mL with distilled water.

   Now add 15 mL 1 N NH_4F and 25 mL of 0.5 N HCl to 460 mL of distilled water. This gives a solution composition of 0.03N NH_4F in 0.025N HCl.

   Or, Dissolve 1.11 g of NH_4F in one litre of 0.025 N HCl. This solution is 0.03N NH_4F + 0.025N HCL

2. *Molybdate reagent* (1.5 % Dickman and Bray's reagent): Dissolve 15 g $(NH_4)_2MoO_4$ in 300 ml of warm distilled water. Add the solution to exact 350 ml of 10 N HCl solution gradually with stirring. Dilute to 1 litre with distilled water.

3. *Stannous chloride stock solution*: Dissolve 10 g $SnCl_2.2H_2O$ in 25 ml of concentrated HCl. Add a piece of pure metallic tin (Zn metal to prevent oxidation) and store the solution in a glass stoppered amber colour bottle.

4. *Stannous chloride working solution*: Dilute 0.5 ml of the stock solution of stannous chloride to 66.0 ml with distilled water just before use.

Primary P standard of 100 ppm (stock solution):

Potassium dihydrogen phosphate (KH_2PO_4) is dried at 40°C in an oven and 0.439g of KH_2PO_4 is dissolved in about half a litre of distilled water. Then 25 mL of 7N H_2SO_4 is added and volume is to 1 litre with distilled water. This gives 100 ppm stock solution of P.

Secondary P standard:

Prepare 2 ppm solution of P by diluting the stock solution of 100 ppm P 50 times.

Procedure:

1. Weigh 5 g of soil sample in a 150 mL of conical flask.
2. Add 50 mL of the extractant and shake for 5 minutes on a mechanical shaker.
3. Filter the solution through Whatman No.1 filter paper to filter within 10 minutes.
4. Transfer 5 mL of the aliquot into a 25 mL volumetric flask.
5. Add 5 mL of ammonium molybdate solution shake a little and dilute to about 22 mL.
6. Add 1 mL of $SnCl_2$ working solution. Mix by shaking a little and make up the volume.
7. Prepare a blank following all the above steps without taking soil.
8. Measure the intensity of the blue colour developed, at 660 nm wavelength (red filter) by a UV-VIS spectrophotometer.

Preparation of standard curve for P:

1. Take 25 mL volumetric flasks.
2. Pipette 0, 1.0, 2.0, 3.0, 4.0 and 5.0 mL of 2 ppm P working standard to prepare P-standard of 0.08, 0.16, 0.24, 0.32 and 0.40 ppm respectively.
3. Add 5 mL of the extractant.
4. Add ammonium molybdate and proceed for the development of blue colour as described in the procedure.
5. Measure the intensity of blue colour and draw a standard curve by plotting concentrations of P in ppm against the absorbance reading.
6. Find out the concentration of P in the test solution from this calibration curve. An example of standard curve is given below for reference.

Observations & Calculations

a.	Weight of soil taken	: 5 g
b.	Volume of the extract added	: 50 mL
c.	Volume of the extract taken for colour development	: 5 mL
d.	Final volume made with colour development (test solution)	: 25 mL
e.	Dilution	: 25/5 =5
f.	Concentration of P in the test solution as read from standard curve	: C ppm
g.	Concentration of P in the original extract	: 5C ppm
h.	Amount of P in 50 mL of the extract	: 5C x 50 µg
i.	Amount of P in 5 g soil	: 5C x 50 µg
j.	Amount of P in 1g soil	: C x 50 x 10^{-6}g
k.	Amount of P in 1 kg soil	: C x 50 x 10^{-6} kg
l.	Amount of P in 1 ha (2.24 x 10^{-6} kg) of soil	: C x 50 x 2.24 kg
	So, Available P (Kg/ha)	: C x 50 x 2.24
	Available P_2O_5(Kg/ha)	: C x 50 x 2.24 x 2.29

3.12.2 Determination of available P in soil by Olsen's method

In this method, the soil is shaken with 0.5 M $NaHCO_3$ (pH 8.5) in 1:20 soil to solution ratio, for half an hour, in the presence of Darco-G 60 (which absorbs the dispersed organic matter in soil and helps in giving a clear extract). The extract is treated with ammonium molybdate (a complexing agent) to form a heteropoly complex, known as phosphomolybdate (yellow coloured). Phosphomolybdate is then reduced by $SnCl_2$ to produce a characteristic blue colour.

Olsen's method to determine available P is followed for neutral, alkaline, calcareous and rice growing soils. The extractant used is 0.5 M $NaHCO_3$ adjusted to pH 8.5. It extracts non specifically held P and phosphorus from Fe-P and Al-P by precipitating Fe and Al as $Fe(OH)_3$ and $Al(OH)_3$. The changes to when soil pH is > 8.5. Carbonate decreases the activity of Ca and therefore increases the solubility of $Ca_3(PO_4)_2$ or the apatite minerals as per their solubility product constant. Thus some phosphate form the surface of these minerals is extracted to the soil solution.

In Olsen's method, the soil is shaken with 0.5 M $NaHCO_3$ (pH 8.5) for 30 minutes in presence of Darco-G-60 which absorbs the dispersed colloidal particles to give a clear extract. The extract is treated with ammonium molybdate to form a heteropoly complex of phosphomolybdate which is reduced by $SnCl_2$ to produce a characteristic blue colour. The intensity of the blue colour is directly proportional to P-concentration on the extract and is measured at 660 nm wavelength. In the above reduction reaction Mo is reduced from + 6 to +5 or to +3 oxidation states to produce the blue colour.

Equipments/Apparatus required

Same as for Bray's Method No. 1.

Reagents required

1. *Bicarbonate (0.5 M $NaHCO_3$) extractant*: Dissolve 42 g of P free Sodium bicarbonate in 1 litre of distilled water and adjust the pH to 8.5 by addition of dilute NaOH or HCl. Filter, if necessary.
2. *Activated Charcoal*: Darco G 60.
3. *1.5 % Molybdate reagent*: Same as for the Bray's Method No. 1.
4. *Stannous chloride solution*: Same as in Bray's Method No. 1.
5. *Standard curve of P*: Same as in Brays' method No.1.

Procedure

1. Weigh 2 g of soil in a 100 mL conical flask.
2. Add a pinch of Darco-G 60 and 40 mL of 0.5 M $NaHCO_3$ solution.
3. Shake the contents for half an hour by an electrical shaker and filter the suspension through Whatman No.1 filter paper.
4. Prepare a blank following all the above steps except taking the soil.

Colour development

5. The colour is developed following the same procedure as discussed in Brays' method.

Observations & Calculations

a. Weight of soil taken : 2 g

b. Volume of the extractant added : 40 mL

c. Volume of extractant/Wt. of soil : 40/2 = 20

d. Volume of the extractant taken for colour development : 5 mL

e. Final volume made with colour development : 25 mL

f. Dilution factor : 25/5=5

g. $\frac{\textit{Volume of extractant}}{\textit{Wt.of soil}}$ x Dilution factor : 20 x5 = 100

h. Concentration of P read from standard curve : C ppm

i. Available P in soil : C x 100 ppm

j. Available P (Kg/ha) : C x 100 x 2.24

k. Available P_2O_5(Kg/ha) : C x 100 x 2.24 x 2.29

Interpretation of Results

Amount of Available P (Kg/ha)	Comments
<10	Low
10-25	Medium
>25	High

3.12.3 Available P by Ascorbic Acid Method

The major drawback of blue colour development method (Dickman and Bray, 1940) is that the colour starts fading soon and hence the intensity has to be measured quickly. The ascorbic acid method has clean edge over the methods described above, as the reduced blue colour complex, due to ascorbic acid is stable for 24 hours and is not affected by moderate variation in temperature, acidity and reducing agents. However, the blue colour developed by Olsens' methods starts fainting after 15 minutes.

Principle

It is based on the reduction of ammonium molybdate complex by ascorbic acid, in presence of antimony-potassium tartarate. The blue colour is produced which is stable for 24 hours and is measured by spectrophotometer.

Reagents

1. *Bicarbonate (0.5 M $NaHCO_3$) extractant*: Dissolve 42 g of P free Sodium bicarbonate in 1 litre of distilled water and adjust the pH to 8.5 by addition of dilute NaOH or HCl. Filter, if necessary.

2. *Activated Charcoal*: Darco G 60

3. *Molybdate-Tartarate Solution*: Dissolve 12 g of ammonium molybdate in 250 mL of distilled water to get solution A. In 100 mL of distilled water dissolve 0.291 g of antimony-potassium tartarate to get solution B. Add both these solutions to 1000 mL of approximately 5 N H_2SO_4 and add solutions 'A' and 'B'to it. Mix thoroughly and make the volume to 2 litres with distilled water.
4. *Ascorbic acid solution*: Dissolve 1.056 g of ascorbic acid in 200 mL of the molybdate-tartarate solution and mix well. Prepare it fresh as and when required.
5. *p-nitrophenol indicator*: Dissolve 0.5 g of p-nitrophenol in 100 mL of distilled water.
6. *5 N H_2SO_4*
7. *Primary P standards* (as per Bray's method)
8. *Secondary P standards* (as per Bray's method)

Procedure

1. Pipette out 5 mL of the Olsen's reagent into a 25 mL volumetric flask add 2-3 drops of p-nitrophenol which will develop yellow colour.
2. Add known quantity of 5 N H_2SO_4 drop by drop to acidify the Olsen's reagent to pH 5.0 at which the yellow colour will disappear. Note the volume of 5 N H_2SO_4 used.
3. Transfer a 5 mL of aliquot of the Olsen's extract of the soil to a 25 mL of volumetric flask. Add the required quantity of 5 N H_2SO_4 to bring the pH to 5.0.
4. Dilute to 20 mL with the distilled water.
5. Add 4 ml of ascorbic acid solution and make the volume to 25 mL with distilled water and shake it well.
6. Wait for 10 minutes and the reading at 730-840 nm.
7. Run a blank with the extracting solution without soil.
8. Prepare the standard curve as described for the Dickman and Bray's method.

3.13 Available Potassium (K) in Soil

Principle

Available Potassium in soil includes the water soluble and exchangeable potassium. It is extracted by neutral Normal ammonium acetate (pH 7.0) solution. When NH_4OAC is added to soil, NH_4^+ replaces K^+ as well as all other exchangeable cations to the soil solution. The concentration of in 1N NH_4OAC is sufficiently high to increase the rate of direct reaction much higher than the rate of reversed reaction to replace all the exchangeable cations to the soil solution. Potassium is among this exchangeable cation. Potassium in the NH_4OAC extract of soil which contains the exchangeable and water soluble K is determined flame photometrically. Because of the low hydration energy, may remain in dehydrated state even in moist soil condition and causes collapse of the clay layers when occupy the hexagonal oxygen cavity. Thus prevents the release of fixed K by the in the extracting solution.

Reactions

$$\text{Clay}\left|\begin{array}{l}\text{H}\\ \text{Al}\\ \text{Na}\\ \text{K}\\ \text{Ca}\\ \text{Mg}\end{array}\right. + 10\,\overset{+}{NH_4} \rightleftharpoons \text{Clay}\ 10\,\overset{+}{NH_4} + \left|\begin{array}{l}H^+\\ Al^{3+}\\ Na^+\\ K^+\\ Ca2^+\\ Mg2^+\end{array}\right.$$

Apparatus required

Flame Photometer, pH-meter, Electrical Balance, 100 mL conical flask, 25 mL Volumetric pipette, Filter paper

Reagents required

1. *Neutral N NH_4OAc (ammonium acetate) solution*: Dissolve 77.09 g of ammonium acetate ($NH_4C_2H_3O_2$) in 1 litre of water. Check the pH by a pH meter. Adjust the pH to 7.0 by adding ammonia or acetic acid solution.
2. *Primary standard of K (stock solution of K):* Dissolve 1.908 g pure AR grade of KCl in 1 litre of distilled water. This solution contains 1 mg K/ml (1000 ppm).

Working standards of K

Pipette 0, 1, 2, 3 and 4 mL of 1000 ppm K into 100 mL volumetric flasks. Dilute to 100 mL by 1N neutral NH_4OAc solution. This gives the working standards of 0, 10, 20, 30 and 40 ppm of K.

Procedure

1. Weigh 5 g of soil in a 100 mL conical flask.
2. Add 25 mL of neutral N NH_4OAc solution to it.
3. Shake the content of the conical flask by an electric shaker for 5 minutes and filter.
4. Determine the concentration of K in the NH_4OAc extract of the soil from this calibration curve. An example of standard curve is given below for reference.

Preparation of standard curve for K:

Adjust the air and gas pressure at the required level. Then adjust the flame photometer reading to zero for 0 ppm K and 100 for 40 ppm K. Take reading for 10, 20, and 30 ppm K. Then take the flame photometer reading for K in the NH_4OAc extract. Draw the calibration curve by plotting the flame photometer reading in Y axis against different K standards (0, 10, 20, 30 and 40 ppm).

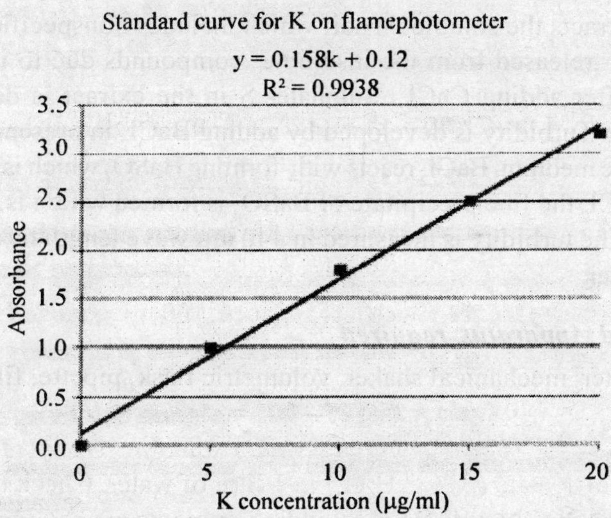

Observations and Calculations

a. Weight of the soil : 5 g

b. Volume of neutral N NH_4OAc solution : 25 mL

c. Flame photometer reading for K in the NH_4OAc extract : R

d. Concentration of K (ppm) as read from the standard curve : C

e. $\frac{\textit{Volume of extracting solution added}}{\textit{Wt. of soil}}$: 25/5 = 5

f. Available K (ppm) in soil : C x 5

g. Available K (kg/ha) in soil : C x 5 x 2.24

h. Available K_2O (kg/ha) in soil : C x 5 x 2.24 x 1.20

Interpretation of Results:

Amount of Available K (Kg/ha)	**Rating**
< 118	Low (Response most likely)
118 - <280	Medium (May or may not respond)
≥280	High (Little or no response)

3.14 Available Sulphur (S) in Soil

3.14.1 Determination of available sulphur (S) by $CaCl_2$ extractant

Principle

Available SO_4^{2-} – S in soil is determined in 0.15 % $CaCl_2$. $2H_2O$ extract of soil. This solution extracts the soluble in soil which includes nonspecifically held and some of the released from the insoluble compounds due to increasing more strength after adding $CaCl_2$. Sulphate- S in the extract is determined turbidimetrically. Turbidity is developed by adding $BaCl_2$ in presence of gum acacia in an acidic medium. $BaCl_2$ reacts with forming $BaSO_4$ which is insoluble. After adding $BaCl_2$ the fine precipitate of $BaSO_4$ is formed which is stabilized by gum acacia. The turbidity is measured in 440 mμ wave length by a UV-VIS spectrophotometer.

Equipments and Apparatus required

Spectrophotometer, mechanical shaker, volumetric flask, pipette, filter paper, conical flask, etc

Reactions:

$CaCl_2 . 2H_2O \rightarrow Ca^{2+} + 2Cl^- + 2H_2O$

Soil Colloid] $SO_4^- + 2Cl^- \rightarrow$ Soil colloid] $2Cl^- + SO_4^{2-}$

$BaCl_2 \rightarrow Ba^{2+} + 2Cl^-$

$Ba^{2+} + SO_4^{2-} \rightarrow BaSO_4$ (Precipitate)

Reagents

1. *0.15 % $CaCl_2$ Solution*: Dissolve 1.5 g of $CaCl_2$. $2H_2O$ in about 500 mL of distilled water and make the volume to 1 litre.

2. *Gum acacia-acetic acid solution (0.25 %):* Dissolve 0.25g of chemically pure gum acacia powder in 100 ml of hot water and filter in hot condition through Whatman No.42 filter paper. Cool and dilute to one litre with dilute acetic acid.

3. *Barium chloride*: Pass AR grade $BaCl_2$ salt through 1 mm sieve and store for use.

4. *Stock solution of S (100 ppm):* Dissolve 0.5434g of oven-dried AR grade potassium sulphate in 1 litre water of distilled water.

Preparation of standard curve for S

Pipette 0.25, 0.5, 1.0, 2.5, 5.0 and 7.5 mL of the stock solution in series of 25 mL volumetric flask. Add to each flask, 10 mL of the extracting solution, followed by about 1 g of $BaCl_2$ crystals and for 1 minute. Then add 1 mL of 0.25 % gum acacia. Make up the volume of each flask with distilled water and shake for 1 minute. This will give the concentration of 1, 2, 4, 10, 20, 30 ppm of S. The reading should be completed within 5-30 minutes at 440 mμ wavelength.

Procedure

1. Weigh 20 g of soil sample in a 250 ml conical flask.

2. Add 100 ml of the 0.15 % $CaCl_2$ extracting solution and shake for 30 minutes. Filter through Whatman No.42 filter paper under suction.

3. Take 20 ml of the clear filtrate into a 25 ml volumetric flask and then proceed further as described in the preparation of standard curve.

Observations and Calculation

a. Weight of the soil taken : 20g

b. Volume of the extractant added : 100 mL

c. Volume of the aliquots taken : 20 mL

d. Final Volume : 25 mL

e. Absorbance as read from the Spectrophotometer : A (say)

f. Available S (ppm) from the standard curve : C (say)

g. $\dfrac{\textit{Volume of extractant added}}{\textit{Wt.of the soil taken}}$: 100/20 = 5

h. $\dfrac{\textit{Final Volume}}{\textit{Volume of aliquot taken}}$: 25/20 = 1.25

i. Total dilution : 5 x 1.25 = 6.25

j. Available S (ppm) in soil : C x 6.25

k. Available S (kg/ha) in soil : C x 6.25 x 2.24

3.14.2 Determination of available sulphur (S) by Monocalcium phosphate extractant

Principle

Soil is shaken with a solution of monocalcium phosphate, containing 500 ppm of P. During this extraction, phosphate ions displace the adsorbed sulphate. Calcium ions depress the extraction of soil organic matter, thus eliminating contamination from extractable organic S. This method extracts soluble SO_4^{2-}, plus a fraction of the adsorbed SO_4^{2-}. The filter is then analysed for S by turbidimetric method.

Turbidimetric procedure

The filtrate is treated with barium chloride in presence of gum acacia solution and the turbidity produced by the precipitation of SO_4^{2-} as barium sulphate is measured spectrophotometrically. Gum acacia, in this determination, helps to prevent too rapid settling of the barium sulphate precipitates.

Reactions

$Ca(H_2PO_4)_2 . 2H_2O \rightarrow Ca^{2+} + 2H_2PO_4^- + 2H_2O$

Soil Colloid] $SO_4^{2-} + 2H_2PO_4^- \rightarrow$ Soil colloid] $2H_2PO_4^- + SO_4^{2-}$

$BaCl_2 \rightarrow Ba^{2+} + 2Cl^-$

$Ba^{2+} + SO_4^{2-} \rightarrow BaSO_4$ (Precipitate)

Equipments and Apparatus required

Spectrophotometer, mechanical shaker, volumetric flask, pipette, Whatman No.42 filter paper, conical flask, suction pump, etc

Reagents required

1. *Monocalcium phosphate extracting solution (500 mg P/litre)*: Dissolve 2.18 g of $Ca(H_2PO_4)_2.H_2O$ in 1 litre of water.

2. *Gum acacia-acetic acid solution (0.25 %)*: Dissolve 0.25g of chemically pure gum acacia powder in 100 ml of hot water and filter in hot condition through Whatman No.42 filter paper. Cool and dilute to one litre with dilute acetic acid.

3. *Barium chloride*: Pass AR grade $BaCl_2$ salt through 1 mm sieve and store for use.

4. *Standard stock solution (2000 mg S/litre)*: Dissolve 1.089 g of oven-dried AR grade potassium sulphate per 100 mL.

5. *Standard working solution (10 mg S/litre)*: Measure exactly 2.5 ml of the stock solution and dilute to 500 ml.

6. *Barium sulphate seed suspension*: Dissolve 18 g of AR grade $BaCl_2$ in 44 ml of hot water and add 0.5 ml of the standard stock solution. Heat the content to boiling and then cool quickly. Add 4 ml of gum acacia-acetic acid solution to it. Prepare a fresh seed suspension for estimation every day.

7. *Dilute nitric acid (approx 25%)*: Dilute 250 ml of AR grade conc. HNO_3 to one litre.

8. *Acetic-phosphoric acid*: Mix 900 ml of AR grade glacial acetic acid with 300 ml of H_3PO_4 (AR grade).

Preparation of standard curve

Pipette 2.5, 5.0, 7.5, 10.0, 12.5 and 15.0 mL of the working standard solution in series of 25 mL volumetric flask to obtain 25, 50, 75, 100, 125 and 150 ppm of S. Proceed to develop turbidity as described in the procedure. Read the turbidity intensity and prepare the curve by plotting readings against sulphur concentrations (in ìg in the final volume of 25 ml).

Procedure

1. Weigh 20 g of soil sample in a 250 ml conical flask.
2. Add 100 ml of the monocalcium phosphate extracting solution (500 mg P/litre) and shake for 30 minutes. Filter through Whatman No.42 filter paper under suction.
3. Take 10 ml of the clear filtrate into a 25 ml volumetric flask and then proceed further as described in the preparation of standard curve.
4. Add 2.5 ml of 25% HNO3 and 2 ml of acetic-phosphoric acid. Dilute to about 22 ml, stopper the flask and shake well, if required.
5. Shake the BaSO4 seed suspension and then add 0.5 ml of it, and 0.2 g of $BaCl_2$ crystals. Stopper the flask and invert three times and keep.
6. After 10 minutes, invert 10 times. Again after 5 minutes invert for 5 times.

7. Allow to stand for 15 minutes and then add 1 ml of gum acacia-acetic acid solution.
8. Make up the volume to 25 ml, invert 3 times and keep aside for 90 minutes.
9. Invert 10 times and measure the turbidity intensity at 440 nm (blue filter).
10. Run a blank side by side.

Observations and Calculation:

a. Weight of the soil taken : 20g

b. Volume of the extractant added : 100 mL

c. Volume of the aliquots taken : 10 mL

d. Final Volume : 25 mL

e. Absorbance as read from the Spectrophotometer : A (say)

f. Available S (ppm) from the standard curve : C (say)

g. $\frac{\textit{Volume of extractant added}}{\textit{Wt.of the soil taken}}$: 100/20 = 5

h. $\frac{\textit{Final Volume}}{\textit{Volume of aliquot taken}}$: 25/10 = 2.5

i. Total dilution : 5 x 2.5 = 12.5

j. Available S (ppm) in soil : C x 12.5

k. Available S (kg/ha) in soil : C x 12.5 x 2.24

Interpretation of Results

Amount of Available S (ppm)	Comments
< 10	Low (Response most likely)
10-30	Medium (May or may not respond)
>30	High (Little or no response)

3.15 Exchangeable Calcium (Ca) & Magnesium (Mg) Content in Soil

Exchangeable Ca^{++} and Mg^{++} are those held on the soil colloidal surface by electrostatic force of attraction. These are in exchangeable forms since they can be replaced by another cation in the soil solution. The exchange reaction follows the laws of mass action.

Principles

Exchangeable Ca^{++} and Mg^{++} are extracted by a neutral N NH_4OAc solution. $Ca^{++} + Mg^{++}$ in the extract is determined by titrating against EDTA using EBT indicator. The analysis is based on the difference in the stability constants of different complexes formed during this experiment.

Reagents

1. *Neutral normal ammonium acetate (CH_3COONH_4) solution*: Dissolve 77.09 g of ammonium acetate in 900 mL of distilled water. Adjust the pH to 7.0 by adding NH_3 solution or acetic acid and makeup the volume to 1 litre by adding distilled water.
2. *Buffer solution*: Dissolve 67.5 g of NH_4Cl in 200 mL of distilled H_2O. Add 570 mL of concentrated NH_4OH and dilute the solution to a volume of 1 liter with distilled H_2O.
3. *Standard EDTA solution*: Dissolve 1.8613 g of disodium salt of ethylenediaminetetraacetic acid (Molecular weight = 372.254 g) in distilled H_2O, and dilute the solution to a volume of 1 litre. This solution is 0.005 M or 0.01 N. The titre of this solution will change with time if stored in glass containers, but not if stored in polyethylene containers.
4. *Standard calcium solution*: Dissolve 0.5004 g of AR grade $CaCO_3$ (dried at 150° C) in 5 mL of approximately 6 N HCl and dilute the solution to a volume of 1 litre. This solution is 0.005 M.
5. *Potassium cyanide solution (1%)*: Dissolve 1 g of potassium cyanide (KCN) in 100 mL of distilled H_2O.
6. *Hydroxylamine hydrochloride solution (5%)*: Dissolve 5 g of hydroxylamine hydrochloride (NH_2OH. HCl) in 100 mL of distilled H_2O.
7. *Potassium ferrocyanide solution (4%)*: Dissolve 4 g of reagent-grade $K_4Fe(CN)_6 .3H_2O$ in 100 mL of distilled H_2O.
8. *Triethanolamine*, reagent grade.
9. *Eriochrome Black T indicator*: Dissolve 0.2 g of EBT in 50 mL of methanol. Prepare a fresh solution every 3 weeks.

Apparatus required

1. Analytical balance
2. 50 mL and 100 mL conical flasks
3. 10 mL and 50 mL volumetric pipettes

4. Reciprocating shaker
5. Burette
6. Filter paper

Procedure

Take 5 g of soil in a 100 mL conical flask. Add 50 mL of 1 N neutral NH_4OAc solution. Shake the contents for ½ hour in a reciprocating shaker. Filter the suspension to another 100 mL conical flask using ordinary filter paper. Discard the first few mL of the filtrate.

Take 10 cc. of the filtrate in a 50 mL conical flask. Add 15 drops of NH_4OH and NH_4Cl buffer solution. Add 10 drops each of KCN, K_4-Fe $(CN)_6$, $NH_2OH.HCl$ (Hydroxylamine hydrochloride) and triethanol amine. Allow a few minutes for the reaction to take place, especially the manganese ferrocyanide reaction. Warming the solution will speed the reaction. Add 10 drops of EBT indicator and titrate against a 0.01 N EDTA till the initial wine red colour of the indicator changes to permanent blue. Run a blank side by side following all the above steps except taking the soil.

Reactions

$Ca^{2+} + EBT \leftrightarrow Ca - EBT^{2+}$ (Wine red)

$Mg^{2+} + EBT \leftrightarrow Mg - EBT^{2+}$ (wine red)

$Ca - EBT^{2+} + EDTA \leftrightarrow Ca - EDTA^{2+} + EBT$ (Permanent Blue)

$Mg - EBT^{2+} + EDTA \leftrightarrow Mg - EDTA^{2+} + EBT$ (Permanent Blue)

In all the above reactions the pH of the NH_4OAc extract is maintained between 7.0 to 11. 0. The free indicator is blue at this pH range, but wine red when it form complex with Ca^{2+} or Mg^{2+}.

Calculation

Exchangeable Ca^{++}+Mg^{++} (meq/100 g)

$$= (S - B) \times N \times \frac{50}{V} \times \frac{100}{w}$$

Where

S = Sample titration value of EDTA (mL)

B = Blank titration value of EDTA (mL)

N = Normality of EDTA (mL)

V = Volume of the extract taken (mL) for titration

w = Weight of soil taken (g)

The above formula is derived as follows

S×N = meq of EDTA consumed in the sample titration

= meq of Ca^{++}+Mg^{++} in V mL of the extract

B×N = meq of EDTA consumed in the blank titration

= meq of Ca^{++}+Mg^{++} in V mL of blank

∴ meq of Ca^{++}+Mg^{++} actually present in V mL of the extract

= (S×N – B×N)

= (S-B) × N

∴ meq of Ca^{++}+Mg^{++} in 50 mL of the soil extract

$$= (S\text{-}B) \times N \times \frac{50}{V}$$

∴ $(S\text{-}B) \times N \times \left[\frac{50}{V}\right]$ meq of exchangeable Ca^{2+}+Mg^{2+} is present in w g of soil.

∴ Exchangeable Ca^{2+}+Mg^{2+} (meq/100 g soil) $= (S\text{-}B) \times N \times \left[\frac{50}{V}\right] \times \frac{100}{w}$

3.16 Available Micronutrient (Fe, Mn, Cu, Zn) Content in Soil

3.16.1 DTPA Extraction Method: (0 .005 M DTPA+0.01 M $CaCl_2$.2 H_2O+ 0.1 M TEA)

Principle

The method consists of shaking a few gram of soil with a buffered solution, containing DTPA (Diethylene tri amine penta acetic acid). DTPA offers the most favourable combination of stability constants for the simultaneous complexing of Zn, Cu, Fe, Mn, Cd, Co, Ni and Pb. This chemical acts as a mild chelating agent and extracts the easily soluble Zn, Cu, Fe and Mn. Chelate reduces the free metal ion activity by forming metal-chelate complex similar to that of absorption of metal by plant from the soil solution. The extracting solution is buffered at pH 7.3 by triethanol amine (TEA) and in addition includes calcium chloride to suppress the dissociation of calcium carbonate to prevent the release of occluded micronutrients. It is a weak base having the pK_b value of 7.8. At the selected pH of 7.3 three fourth of TEA is protonated and is present as $HTEA^+$. When the extractant is added to the soil, additional Ca^{+2} and some Mg^{+2} enter into soil solution, largely because the protonated TEA exchanges with Ca^{+2} and Mg^{+2} from soil exchange sites. This raises the concentration of ionic Ca^{+2} by 2 to 3 fold and aids in suppressing the dissolution of $CaCO_3$. These conditions permit the right amount of micronutrients to be dissolved and $CaCl_2$ is to stabilize the pH of the extractant. The capacity of DTPA to complex each of the micronutrients cations is 10 times its atomic weight and ranges from 550 to 650 ppm depending on the micronutrient cation. Thus, DTPA is present in excess of the micronutrient metal cations that are normally solublised during extraction. This excess reduces the possibility that the extraction of one micronutrient might significantly affect the amount of other micronutrient extracted. The dissolved elements in the extract are then measured by the atomic absorption spectrophotometer, where in the extracted sample is converted first into an atomic vapour by a flame and irradiated by the metal being sought. The absorption of the light by the vaporized samples is related to the concentration of the desired metal in it.

The narrow emission lines which are to be absorbed by the sample are generally provided by a hollow cathode lamp filled with neon or argon at a low pressure, which has a cathode made of the element being sought. Such a lamp emits only the spectrum of desired element together with that of the filler gas. In the atomizer burner system the solution to be analyzed is drawn up by a capillary and converted by means of a stream of compressed air into a fine spray, which after condensation of larger droplets, is mixed with acetylene and burned in a long flame at a stainless steel burner. The light from the lamp, after traversing the flame enters a momochromator, which has been set at a wavelength of the resonance line of the metal to be determined and then falls on the photomultiplier tube. The light is amplified by an A.C amplifier whose output is read on a meter.

Reagents

1. *DTPA Solution*: Dissolve 1.9679 g of AR grade DTPA (Diethylene tri amine penta acetic acid + 13.3 ml reagent grade TEA (Triethanol amine) and 1.47 g $CaCl_2$ $2H_2O$ in 200 ml distilled water, allow sometime for DTPA to dissolve, dilute to 900 ml, adjust pH to 7.3 ± 0.05with 6 N HCl while stirring and then make upto 1 litre and mix thoroughly.
2. Standard Micronutrient Solutions(Primary Stock)
   i. 1000 ppm Zn (4.398 g/L of $ZnSO_4.7H_2O$) or 1 g Zn wire in minimum volume of 1:1 Nitric acid and dilute to 1000 ml or analytical grade

1000 ppm stock solutions directed purchased from chemical stores)

ii. 1000 ppm Cu(3.929 g/L of $CuSO_4.5H_2O$) or 1 g Cu wire in minimum volume of 1:1 Nitric acid and dilute to 1000 ml or analytical grade std solutions can be directly purchased)

iii. 1000 ppm Fe(4.977 g/L of $FeSO_4.7H_2O$) or 1 g Fe wire in minimum volume of 1:1 Nitric acid and dilute to 1000 ml or analytical grade std solutions can be directly purchased)

iv. 1000 ppm Mn(3.598 g/L of $MnSO_4.H_2O$) or 1 g Mn wire in minimum volume of 1:1 Nitric acid and dilute to 1000 ml or analytical grade std solutions can be directly purchased)

3. *Secondary micronutrient standards* . 100 ppm (dilute 10 ml of 1000 ppm stock to 100 ml with double distilled water)
4. *Working Micronutrient standards*: Series of Micronutrient standards can be preparred from secondary standard in the range of 0-5 ppm for Fe,Mn,Cu (1, 2.5, 5 ppm) and 0-2 ppm for Zn

Apparatus Required

1. Shaker(Horizontal/Rotary)
2. Conical flasks with glass stoppers(100 ml)
3. Funnel
4. Filter paper Whatman No 42
5. Plastic storage bottles
6. AAS
7. Acetylene gas

Procedure

Soil Extraction: Weigh 10 gm of air dried soil sample in a 125 ml conical flasks or polypropylene bottle. Add 20 ml of DTPA solution in the ratio of 1:2. Cork the bottles or flask and place them upright on a shaker .Shake the mixer for 2 hours on shaker at 70-80 oscillations per minute at 25 ±1 °C. Filter the suspension through whatman no 42 filter paper and collect the filtrate in plastic bottles. Then filtrate may be directly fed to AAS or after necessary dilution to determine the content of micronutrients after standardization with respective standards.

Calculation

DTPA Micronutrient (ppm or mg/kg) =

AAS reading (ppm) x 2 (Solution/soil) x dilution factor

Precautions:

1. The DTPA test is a nonequilibrium extraction. Therefore factors viz. shaking time, shaking speed and shape of extracting vessel influence the quantity of metal extracted. These factors must be standardized in each laboratory
2. For determination of Fe and Mn 10 to 20 times dilution is required after filtration
3. While taking reading for Zn & Cu dilute HNO_3(10%) solution may be sucked time to time to keep the burner carbon free for better atomization

Rating: It is the soil status below which status is deficient and needs fertilization with respect to micronutrients

Element	Critical limits (mg/kg)
Zn	0.6
Cu	0.2
Fe	4.5
Mn	3.5

3.17 Available Boron (B) Content in Soil

3.17.1 Determination of Hot water soluble Boron in soils by Azomethine-H colorimetric Method

Boron (B) is unique in soils by its narrow range between deficiency and toxicity. Less than 1 ppm of B may lead to a deficiency, yet 3 ppm may be toxic. It is associated with calcium uptake by plants and in fertility studies it is often useful to measure boron-calcium ratio. The most useful measure of available B is the water soluble forms. The content of water soluble B in soils is influenced by pH, organic matter and amount of colloids.

Once extracted from the soil, B can be analyzed by colorimetric methods using different reagents. Of the numerous reagents used for colorimetric determination of B, quinalizarine, curcumin and azomethine-H are the most popular.

Principle

Soil is extracted with boiling water for hot water soluble B, using charcoal for nullifying the effect of organic matter on B adsorption and the clear filtrate is treated with EDTA and colour developing azomethine- H solution in the presence of buffer. The intensity of coloured complex is measured at 470 nm and the concentration of the unknown sample estimated from the standard curve. Azomethine H forms coloured complex with H_3BO_3 in aqueous media. Over a concentration range of 0.5 to 10 µgB/ml the complex is stable at pH 5.1. Maximum absorbance occurs at 420 nm with little or no interference from a wide variety of salts. This technique is rapid, reliable and more convenient to use than traditional procedures.

Apparatus required

1. Spectrophotometer
2. Polypropylene tubes
3. Reflux condensers
4. Water bath/Hot plate

Reagents required

1. Deionised water
2. Buffer solution: Dissolve 250 g of Amonium acetate (NH4OAc) and 15 g of ethylene diamine tetra acetic acid (EDTA disodium salt) in about 400 ml distilled water. Slowly add 125 ml of glacial acetic acid and mix .EDTA acts as a masking agent to suppress Fe from interference
3. Azomethine-H: Dissolve 0.5 gm of Azomethin –H in 1 % L Ascorbic acid solution. Prepare fresh azomethine-H for everyday use.
4. Standard B solution : 100 ppm B : Dissolve 0.570 g of boric acid AR grade in 1 lt of distilled water
5. Prepare working standards ranging from 0-5 ppm B by necessary dilution

Procedure

Weigh 20 gm of soil to a low B pyrex corning or quartz conical flask, add 40 ml of distilled water (1:2 ratio). Add 0.4 gm of charcoal. Boil or reflux the contents for 5 mins over a hotplate by closing the mouth of the conical flasks with reflux condenser tubes. Run a blank taking without soil. Filter the contents by Whatman No 42 filter paper to get a clear filtrate. Filtration should be done while hot. Then take 1 ml of filtrate, 2 ml buffer, mix it thoroughly and 2 ml Azomethine-H in a polypropelene tube, mix the contents and leave for 30 minutes for color development. After 30 mins take the reading of samples along with blank in a spectrophotometer at 420 nm wavelength after standardizing the instrument with a series of B standards 0 - 5 ppm range prepared in a similar manner as that of sample.

Calculation

Wt of soil = w gm

Volume of the extract (hot water) = v ml

B (ppm or mg/kg) = B concentration from standard curve X dilution factor (v/w)

3.17.2 Determination of Extractable Boron in soils by Salicyclic Method for Acid Soils

Hot water extraction of B is a commonly used method for obtaining an index of plant available B in soils. Extracting the soil with boiling water posed problems due to colour from organic matter and turbidity from suspended fine clay particles in the colorimetric estimation of B. Apart from these problems, this method is also difficult to adopt for the analysis of large numbers of samples.

This extractants efficiency in acid soil has not been tested. Similarly there is no evidence to suggest that dilute phosphoric acid is useful as a predictive tool for plant response to B. Although hydro carboxylic acid, such as tartaric acid and salicylic acid are known to react with H_3BO_3, there is little evidence concerning the potentiality of these compounds as extractant in acid soils. Salicylic acid gives an amethyst colour with Fe in solution. It may be inferred that Fe extracted from soils in salicylic acid extracts may produce same colour which may be carried out so far.

It is of considerable interest to find out the relationship of available B with soil characteristics for the use in soil fertility evaluation programme. In general, sandy soils contain less available B than heavy textured soils. The plant availability of B in soils increased up to pH 5.0 and then the availability remained constant up to pH 7.0; above this pH range, B became less available to plant.

Organic matter plays an important role in minimizing B leaching from acid soil and maintaining B in relatively available form. In some acid soils Fe_2O_3 was found to be negatively correlated with extractable B while the influence of $Al(OH)_3$ on extractable B was non-significant. Most of the above studies were related to hot water extractable B only. Information regarding the influence of soil properties on other extractable forms of B in meager.

Principle

The Azomethine-H Method involves the coupling of H-acid with an aromatic hydroxyaldehyde, such as salicylaldehyde, due to the catalytic effect when boron is present. At neutral pH values and a controlled temperature, the condensation reaction is completed quickly (within 15 minutes). After product formation, the solution is adjusted to an acidic pH for optimum color measurement at 420 nm (yellow) using a colorimeter or spectrophotometer. The method is sensitive and highly selective for the determination of dissolved boron in water.

Apparatus required

1. Spectrophotometer
2. Polypropylene tubes
3. Shaking Mechine

Reagents required:

1. *Salicylic Acid* 0.1 M in 10 ml of triethanol amine
2. *Buffer solution*: Dissolve 250 g of Amonium acetate (NH4OAc) and 30 g of ethylenedia aminetetraacetic acid (EDTA disodium salt) in about 400 ml distilled water. Slowly add 125 ml of glacial acetic acid and mix EDTA that acts as a masking agent to suppress Fe from interference. A higher concentration of EDTA compare to 15 gm as describe by John etal (1975) is used in preparation of buffer solution to make the extraction colour free and to check the interference of iron from complex in salicyclic acid.
3. *Azomethin-H*: Dissolve 0.5 gm of Azomethin –H in 1 % L Ascorbic acid solution. Prepare fresh azomethin-H for everyday use.
4. *Standard B solution* : 100 ppm B : Dissolve 0.570 g of boric acid AR grade in 1 lt of distilled water
5. Prepare working standards ranging from 0-5 ppm B by necessary dilution

Procedure

Weigh 10 gm of soil to a low B pyrex corning or quartz conical flask, add 20 ml of 0.1 M salicylic acid (1:2 ratio). Shake the content for 1 hour on end over end shaker. Run a blank taking without soil. Filter the contents by Whatman No 42 filter paper to get a clear filtrate. Then take 1 ml of filtrate, 2 ml buffer and 2 ml Azomethin H in a polypropelene tube, mix the contents and leave for 30 mintutes for color development. After 30 mints take the reading of samples along with blank in a spectrophotometer at 420 nm wavelength after stand riding the instrument with a series of B standards 0-5 ppm range .prepared in a similar manner as that of sample

Calculation

Wt of soil = w gm

Volume of the extract (hot water)= vml

B (ppm or mg/kg) = B conc from standard curve X dilution factor (v/w)

Precautions

Throughout the B analysis, use of borosilicate glassware's should be avoided even for storage of chemicals. Plastic containers or corning pyrex glasswares should be used.

3.18 Cation Exchange Capacity (CEC)

The capacity of negatively charged clays and organic matter to adsorb cations simple physical attractive forces is called the cation exchange capacity (CEC) of the soil. It is seldom that a cation in a soil exists wholly or even largely in the exchangeable form, but the exchangeable form is the most important source of immediately available plant nutrients. Available cations can be considered as exchangeable cations. It is by virtue of this property of soil colloids, that a soil is able to sustain crop growth by regulating plant nutrient supply. As a plant takes up nutrients from the soil solution, the latter is depleted of its nutrient intensity, resulting in replenishment of cations from exchange sites and so a steady supply of nutrients is maintained.

In neutral soils, the major exchangeable cations are Ca^{2+}, Mg^{2+}, K^+, and Na^+. These are referred to as basic a cations. They approximately constitute 80 % Ca, 15 % Mg, 5% (Na + K) of the total exchange capacity of the soil. The proportion of NH_4^+ is of minor occurrence and variable depending upon the nature and amount of organic matter as well as the efficiency of microbial nitrification. Other cations (Mn^{2+}, Zn^{2+}, Cu^{2+}, Fe^{2+}, etc) are held by stronger adsorption forces, on the surface of mineral grains or form chelates with humic acid. A variable amount of H^+ ion is also held on exchange sites.

It can be expressed in terms of milliequivalents/100 g of soil (me/100 g) or centimoles of positive charge per kg of soil (C mol (+) / kg), which is numerically equal to me/100 g. The CEC of the soil depends on the kind of clay, pH, type of ion, concentration, temperature and organic matter present.

Importance of CEC

- The CEC of a soil, together with exchangeable bases, provides a measure for evaluating the fertility status of the soil.
- The determination of the CEC is of considerable significance in studies done for the evaluation of adsorption of cations.
- The CEC determination with exchangeable bases is expected to provide more authentic results on lime requirement of a soil.

Methods of determination

Two methods of CEC determination are discussed in this book.

3.18.1 Distillation Method (Ammonium saturation method; N NH_4OAc, pH 7.0):

Ammonium is used in the determination of CEC because;

- It is the most preferably adsorbed cation among the cations in the lyotropic series.
- It is used in concentrated solutions for efficient and rapid displacement, the high solubility and cheapness of all ammonium salts is relevant.
- The pH dependence of the negative charge in soil colloids can successfully be estimated by NH_4OAc, the salt of a weak acid and a weak base at a pH most suited to the investigation.
- Ammonium salt interfere least in flame and other photometric studies, they are relatively volatile and can be easily removed from soil extracts.

Principle

In this method the CEC of the soil is measured by leaching it with neutral N NH_4OAc solution which saturates the soil surface with NH_4^+ ions, this is followed by the removal of excess salt with alcohol. The adsorbed NH_4^+ is finally determined by distilling the soil with magnesia (MgO) and the gas, evolved during the distillation, is absorbed in a known volume of boric acid taken in excess, the un-reacted acid molecules, being titrated back with a standard acid.

Reaction

$CH_3COONH_4 \leftrightarrow NH_4^+ + CH_3COO^-$

X- Soil + $NH_4^+ + CH_3COO^- \leftrightarrow NH_4$-Soil + CH_3COOX [X may be any cation]

$MgO + H_2O \rightarrow Mg(OH)_2 \rightarrow Mg^{2+} + 2OH^-$

$2NH_4$ – Soil + $Mg^{2+} + 2OH^- \rightarrow 2NH_3 \uparrow$ + Mg-Soil + $2H_2O$

Apparatus required

Buchner funnel, conical flask, distillation set, pipette, filter paper (Whatman No.42), etc.

Reagents

1. *Ammonium acetate (NH_4OAc) 1 N (pH 7.0)*: Dissolve 114 mL of glacial ammonium acetate with water to a volume of approx. one litre. Then add 138 mL of NH_4OH and add water to obtain a volume of about 1.980 mL. Adjust pH to 7.0 with dilute ammonium hydroxide or acetic acid as required and make up the volume to 2 litres.

2. Isopropyl Alcohol (99%) or ethanol (95%)

3. *2% Boric acid*: Dissolve 20 g of boric acid powder in warm water by stirring and dilute to one litre.

4. *Mixed Indicator*: Dissolve 0.066 g of methyl red and 0.099 g of bromocresol green in 100 ml of ethyl alcohol. Add 20 ml of this mixed indicator to each litre of 2% boric acid solution. Adjust the pH to 4.5 with dilute NaOH.

5. 0.02 N H_2SO_4

6. 0.1 N NaOH

Procedure

1. Weigh 10 g of soil sample in a 500 mL conical flask.
2. Add 50 mL of NH_4OAc solution to the soil sample, mix thoroughly and let it stand for overnight.

3. Transfer the contents of the beaker onto a moist filter paper, seated inside a Buchner funnel under suction.
4. Continue to leach the soil with NH_4OAc solution, pouring 25 mL at a time, using in all 150 mL (including 50 mL used earlier). Keep the filtrate for the determination of total and individual exchangeable bases.

5. To the soil on the filter paper, add a pinch of NH_4Cl crystal and leach with alcohol. Continue washing of the residue till the filtrate runs free of chloride (test the filtrate with $AgNO_3$ solution- white turbidity indicates incomplete washing).

6. Transfer the residue and filter paper to a 500 mL distillation flask. Add 0.5 g of MgO and 200-300 mL of distilled water.

7. Connect the flask to the distillation apparatus. Place a 250 mL conical flask containing 25 mL of 2% boric acid with mixed indicator under the condenser of the distillation apparatus to receive NH_3.
8. Start distillation by heating the distillation flask. Continue heating till no NH_3 is evolved through the end of the condenser, which is confirmed by the turning of red litmus paper to blue.
9. Titrate the distillate against 0.02N H_2SO_4 taken in a burette until pink colour starts appearing.
10. One reagent blank should be run with the same volume of NH_4OAc.

Observations and Calculations

a. Weight of soil taken : 10 g

b. Volume of 0.02 N H_2SO_4 required for sample titration : S mL

c. Volume of 0.02 N H_2SO_4 required for blank titration : B mL

d. Meq of 0.02 N H_2SO_4 required for sample titration : S x 0.02 mL

e. Meq of 0.02 N H_2SO_4 required for blank titration : B x 0.02 mL

f. Meq of boric acid consumed for NH_3 absorption : (B-S) x 0.02 mL

g. Now 10 g soil contains of cations : (B-S) x 0.02 meq

h. So, 100 g soil will contain $= \frac{(B-S)\times 0.02\times 100}{10}$ meq of cations

i. Hence, CEC of the soil (meq/100g) $= \frac{(B-S)\times 0.02\times 100}{10}$

3.18.2 Centrifuge Method (Sodium saturation method; N NaOAc, pH 8.2)

Principle

The soil is leached with a sodium acetate solution (pH 8.2) for replacement of exchangeable cations by Na^+ ions. The excess of salts is washed down by alcohol and the adsorbed Na^+ ions are replaced by NH_4^+ ions using N NH_4OAc (pH 7.0) solution. The Na^+ ions, so released, from the exchange spots are measured flame-photometrically and serve as the measure of total exchange capacity of the soil.

Reaction:

$CH_3COONa \leftrightarrow Na^+ + CH_3COO^-$

X- Soil + Na^+ + CH_3COO^- $\leftrightarrow$ Na-Soil + CH_3COOX [X may be any cation]

$CH_3COONH_4 \leftrightarrow NH_4^+ + CH_3COO^-$

Na – Soil + NH_4^+ $\leftrightarrow$ NH_4^+ - Soil + Na^+

Apparatus

Centrifuge; centrifuge tubes (50 or 100 mL), volumetric flask, funnel, mechanical shaker, Flame Photometer and accessories which include Propane, Sodium standards

Reagents

1. *N NaOAc (CH_3COONa), pH 8.2*: Dissolve 136 g of sodium acetate trihydrate, $NaC_2H_3O_2.3H_2O$ in water and dilute to one litre. The pH of the solution should be 8.2. if needed, a few drops of acetic acid or NaOH solution to bring the reaction of the solution to pH 8.2.Add a pinch of NaCl crystal into it.
2. *Ammonium acetate (NH_4OAc) 1 N (pH 7.0)*: Dissolve 114 mL of glacial ammonium acetate with water to a volume of approx. one litre. Then add 138 mL of NH_4OH and add water to obtain a volume of about 1.980 mL. Adjust pH to 7.0 with dilute ammonium hydroxide or acetic acid as required and make up the volume to 2 litres.
3. Isopropyl Alcohol (99%) or ethanol (95%)
4. *Standard solution of NaCl:* Dissolve 5.845 g of AR grade NaCl in 1N ammonium acetate and make the volume to 1 litre. It will give 100 meq/ litre of sodium in stock solution. From this solution take 0, 1, 2, 5, 7.5 and 10 ml and make up the volume to 100 ml each with the ammonium acetate. It will give 0, 1, 2, 5, 7.5 and 10 meq/litre of sodium.

Procedure

1. Weigh accurately 4 g soil and transfer the sample to a 50 ml centrifuge tube in duplicate.
2. Pour 33 mL of N sodium acetate (NaOAc) solution to the tubes, stopper and shake in a mechanical shaker for 5 minutes.
3. Centrifuge at 8000 rpm for 10 minutes or until the supernatant liquid is clear.

4. Decant the liquid completely and repeat the extraction three more times. Discard the decants.
5. Repeat the extraction three more times with 33 mL of NaOAc solution, vertexing each time to ensure uniform suspension of soil particles in the extracting solution.
6. Repeat steps 2 – 4 with ethanol or isopropyl alcohol, shake for 5 minutes, centrifuge and discard the supernatant, continue until the EC of the decant reads less than 40 dS/m (usually it takes 4 to 5 washings) or the test for excess salts with $AgNO_3$ solution, if possible.
7. Replace the adsorbed Na with three 33 mL of N NH_4OAc solution, centrifuging, vortexing and decanting each washing into a 100 mL volumetric flask.
8. Dilute the solution to volume with NH_4OAc solution.
9. Determine the Na concentration in the extract on a flame photometer.
10. Prepare a series of Na standard solutions in the range of 0 – 10 meq/litre of Na. Prepare a standard curve by plotting sodium concentration on x-axis and flame photometric readings on y-axis. Unknown sample extract is fed on the flame photometer and the reading is taken, corresponding to which the concentration of sodium is read from the standard curve. For better results, add LiCl in each standard to yield a final concentration of about 5 meq/litre of LiCl. [10 or 20 ppm of Na strength should be set to 100 reading of the flame photometer].

Observations and Calculations

a. Weight of the soil : W g (say)

b. Volume of NH_4OAc extract prepared : 100 mL

c. Dilution factor : F (say)

d. Flame photometer reading of the sample : G (say)

e. Na concentration read from the standard curve : S ppm (say)

f. Amount of Na in 100 mL of the extract : S x F ppm

Now, 1000 mL of extract contains : S x F mg of Na

So, 100 mL of extract contains : $\frac{S \times F \times 100}{1000}$ mg of Na

Thus, W g of soil contains : $\frac{S \times F \times 100}{1000}$ mg of Na

So, 100 g of soil contains : $\frac{S \times F \times 100 \times 100}{1000 \times W} = $ Y mg of Na

Then, Y mg of Na/ 100 g of soil : $\frac{Y}{23}$ meq. of Na /100 g of soil

Hence, CEC of the soil (meq/100 g) : $\frac{Y}{23}$

3.19 Lime Requirement in Soil

The lime requirement of a soil can be determined by several different methods. Titrating the soil with an acid or base will subsequently decrease or increase the soil pH. The base usually employed is $Ca(OH)_2$ and the acid is HCl. After equilibrium, the pH is determined and the values are plotted against the meq of acid or base added. From these data it is simple to determine the amount of lime to be added.

The methods in use are based on the change in pH of a buffered solution added to soil. When an acid is added to a buffered solution, the buffer solution is depressed in proportion to original soil pH and it's BC. By calibrating pH changes in the buffered solution, the amount of lime required to increase pH to the desired level is called lime requirement which can be calculated.

3.19.1 Determination of lime requirement by Woodruff's buffer method in soil

Crop yields are normally high in soils with pH values between 6.0 and 7.5. Low pH in acid soils declines the crop production. To enhance the crop growth liming is required. The amount of lime (pure $CaCO_3$) required to be added to acidic soil to raise the pH of that soil to a desired value is known as lime requirement. It is the quantity of pure $CaCO_3$ to be added in a soil to raise the pH to a desired level for optimum crop growth. Lime requirement of the soil depends on its buffering capacity where as buffering capacity of soil depends upon its Clay, free oxides of Fe and Al and organic contents. Of the various methods developed to determine the lime requirement of soil, Woodruff's buffer method is still followed because of its simplicity and rapidity. The buffer is prepared with p-nitrophenol, salicylic acid, calcium acetate, calcium hydroxide and the pH is adjusted to 7.0. On adding buffer to the soil, the pH of the buffer decrease below 7.0. Each 0.1 unit decrease in the pH of the buffer corresponds to 1.0 meq of acidity per 100 gm soil. The lime requirement of the soil is calculated on the basis of meq of acidity present in soil.

Reagents

1. *Woodruff's buffer solution:* Dissolve 10 g calcium acetate [$Ca(CH_3COOH)_2$], 12 g p-nitrophenol, 10 g salicylic acid and 4 g calcium hydroxide in 900 mL of distilled water. Adjust pH to 7.0 with dilute acetic acid (CH_3COOH) or sodium hydroxide (NaOH), transfer to 1 L volumetric flask and make the volume to the mark with distilled water.
2. *1 M $CaCl_2$ Solution*: Dissolve 14.7 g of $CaCl_2.2H_2O$ in distilled water and dilute to 100 mL.

Requisites:

pH meter, Polythene beaker, glass rod, Automatic pipettes

Procedure:

1. Take 10 g soil sample in a clean 50 ml polythene beaker.
2. Add 10 ml of distilled water, stir the content by glass rod for 2 to 3 minutes and add 2 drops of 1M $CaCl_2$ solution and wait for 30 minutes.
3. Determine pH value in soil suspension.
4. If the pH value is less than 5.0 (average of 4.5 and 5.5 to have one value), add 10 ml Woodruff's buffer solution.
5. Stir the content with glass rod and determine new pH value after 30 minutes. This is buffer pH of the soil (pH_{sb}).

Calculation

LR i.e. CaCO3 to be added in $kgha^{-1}$ of soil to raise the pH to 7.0

$= (7.0 - pH_{sb}) \times 10{,}000$

LR^1 i.e. CaCO3 to be added in $kgha^{-1}$ to attain a desired pH(pH_s^1)

$$= \frac{LR \, x \, pH \, 7.0}{7.0 - pH_s}\left(pH_s^1 - pH_s\right)$$

Liming material to be added = LR^1 x 100/NV

Where; NV = Neutralizing value or $CaCO_3$ equivalent of the liming material.

3.19.2 Determination of lime requirement by simplified $CaCl_2$-$Ca(OH)_2$ equilibration method in soil

Determination of lime requirement by the simplified $CaCl_2$-$Ca(OH)_2$ equilibration method has recently been developed by Professor Dr. P. K. Das, Department of Soil Science & Agricultural Chemistry, OUAT, Bhubaneswar (unpublished method). Lime requirement determined by this is less than that determined by the modified Woodruff buffer method to attain a desired pH. Modified Woodruff buffer method, which is presently followed, overestimates the LR to a desired pH. The new method also overestimates the LR to the desired pH but the extent of overestimation is higher in modified Woodruff buffer method than this method. The new method is simple and less expensive as compared to the modified Woodruff buffer method. The cost of chemicals required for determining LR by the new method is 20 times less than that for determining LR by modified Woodruff buffer method. To determine LR by the new method, exchangeable H^+ in soil is directly neutralized which is the real neutralization process that occurs in soil after liming. In modified Woodruff buffer method the exchangeable H^+ in soil is only inactivated. A 0.05 molar $CaCl_2$ is used in this method to replace all the exchangeable Al^{3+} for neutralization.

Chemicals Required

1. 1.25 M $CaCl_2$
2. Saturated Ca $(OH)_2$ solution
3. K.H. phthalate

Apparatus Required

pH meter

Procedure

1. Boil 500 mL of distilled water.
2. Add 2.5 mL of Ca $(OH)_2$ to the boiling water and stirre the contents.
3. Cool the solution and filter to a 500 mL volumetric flask using ordinary filter paper. Do not make volume adjustment.
4. Determine the strength of the saturated $Ca(OH)_2$ solution by titrating against a 0.02N KHpthalate. (Take 5 cc of the filtrate in a 25 mL or 50 mL conical flask. Add 2-3 drops of methyl red indicator and titrate against 0.02N KHpthalate till the colour changes from straw yellow to red.)
5. Pipette 10 mL of 1.25 M $CaCl_2$ solution to a 25 mL volumetric flask. Adjust the volume to 25 mL by adding distilled water. This is solution 'A' which is 0.5 M $CaCl_2$ without containing $Ca(OH)_2$.

6. Similarly to another 25 mL volumetric flask add 10 mL of 1.25 M $CaCl_2$ solution and 5 mL of saturated $Ca(OH)_2$ solution. Make the volume to 25 mL by adding distilled water. This is solution 'B' which is 0.5 M $CaCl_2$ containing $Ca(OH)_2$.
7. Take 10g soil each in two separate 50 mL plastic beaker. Add solution A to one beaker and solution B to another beaker. Stirre the contents by a glass rod and measure the pH of the soil suspension after 30 minutes.

Precautions

1. The saturated $Ca(OH)_2$ solution must be used within 3 days of its preparation. It cannot be used after 3 days.
2. During preparation of $Ca(OH)_2$ solution it should not be cooled for long time. Wait till the solution is just cooled so that it can be hold by hand and then immediately start filtration. Care should be taken to obtain a clear filtrate. Close the mouth of the volumetric flask by giving a lid immediately after the filtration is over.
3. The saturated $Ca(OH)_2$ solution should not be immediately exposed to air. The solution left after its use shou!d be transferred to the main container and the lid should be given.

Calculation

LR to attain the desired pH (pH_d)

$$= 8N \times \frac{pH_d - pH_1}{pH_2 - pH_1} \times 1000$$

Where;

N = Normality of saturated $Ca(OH)_2$ solution

pH_1 = pH of the soil suspension after adding $CaCl_2$ solution without containing $Ca(OH)_2$.

pH_2 = pH of the soil suspension after adding 0.5 M $CaCl_2$ solution containing $Ca(OH)_2$.

3.20 Gypsum Requirement of Sodic Soils

In sodic and saline-sodic soil exchangeable Na must be reduced which can be difficult because the soil clay may be dispersed, preventing infiltration. Exchange of Na must be accomplished with Ca by adding the appropriate rate of gypsum ($CaSO_4.2H_2O$).

3.20.1 Determination of gypsum requirement in sodic soils

Principle

Alkali soils contain excessive amounts of sodium and alkalinity (CO_3^{2-} & HCO_3^-). Their soil Na saturation is sufficiently high to interfere with the growth of several crops. To improve their productivity, these soils have to be reclaimed with the immediate objectives of neutralization of exces ive alkalinity and replacement of Na on the exchange complex with Ca in order to improve the physico chemical and biological properties of the alkali soils. These objectives can be achieved by suitable amendment. Gypsum, being the cheapest and easily available, is commonly used as reclaiming material. The gypsum requirement of soils is determined by using the method proposed by Schoonover (1952).

Apparatus required

1. Conical flask,
2. Burette and stand,
3. Pipette,
4. Filter paper,
5. China dish,
6. Glass rod and
7. Mechanical shaker

Reagents/Chemicals required

1. *Saturated gypsum solution*: Add 5 g of $CaSO_4.2H_2O$ to 1 L of distilled water. Shake manually several times over a period of 1 hr or 10 minutes on a mechanical shaker and filter.
2. *Eriochrome Black T indicator*: Dissolve 0.5 g of Eriochrome Black T indicator and 4.5 g hydroxyl amine hydrochloride in 100 mL of 95 % ethanol.
3. *$NH_4Cl - NH_4OH$ buffer*: Dissolve 67.5 g of NH4Cl in 570 mL NH4OH and dilute to 1 L.
4. *EDTA solution (0.01 N)*: Dissolve 2 g of di sodium salt of ethylene diamine tetra acetate and dilute to 1 L. Standardise this solution against 0.01 N calcium chloride solution.

Gypsum reaction in the soil

A known weight of the soil is shaken with a known excess volume of nearly saturated gypsum solution. The unutilised calcium left out in the solution is determined by titrating against standard EDTA solution using EBT indicator in the presence of $NH_4Cl - NH_4OH$ buffer. The amount of gypsum utilised / required is calculated from the titre value.

Soil Na + $CaSO_4 \rightarrow$ Soil Ca + Na_2SO_4 + unutilized gypsum

Micelle Gypsum (saturated)　　　　　　　Sodium in the sulphate extract

Procedure

1. Weigh 4 g of air dried soil into a 250 mL Erlenmeyer flask.
2. Add 100 ml of saturated gypsum solution to remove the Na from exchangeable complex to bring down to a reasonable level.
3. Shake for 5 minutes on mechanical shaker or by hand for several times for 30 min
4. Filter the contents with Whatman No.1 filter paper.
5. Take 5 ml of extract into a china dish.
6. Dilute to approximately to 25 mL with distilled water.
7. Add 1 ml of $NH_4Cl - NH_4OH$ buffer solution followed by 2 to 3 drops of EBT indicator.
8. Titrate against the 0.01 N EDTA solution till the color changes from purple to blue.
9. The titre value will give the volume of the EDTA required to react with the unutilized gypsum.
10. Also run a blank titration with 5 mL of saturated gypsum solution to know the blank titre value.

Observations

S.No.	Aliquot taken	Burette readings		Total Volume (mL) Final – Initial
		Initial (mL)	Final (mL)	
Blank	5 mL			B
Sample	5 mL			S

Blank title value = *B* mL of 0.01 N EDTA

Sample title value = *S* mL of 0.01 N EDTA

Actual titre value = B - S ml

Calculations

a. Wt of the soil taken = 4 g

b. Volume of saturated gypsum added = 100 ml

c. Volume of 0.01 N EDTA required to react with 5 ml of Gypsum solution = **B** ml of 0.01 N EDTA

d. Volume of EDTA required to react with unutilized gypsum solution = **S** ml of 0.01 N EDTA

e. Volume of EDTA required to react with utilized gypsum = **(B-S)** ml 0.01 N EDTA

As per the law of equilibrium;

1000 ml of IN EDTA = 86 g of gypsum

(B-S) ml of 0.01 N EDTA = $\frac{(B-S)ml \times 0.01}{1000} \times 86\, g$ of gypsum

5 ml of extract requires = $\frac{(B-S)ml \times 0.01}{1000} \times 86\, g$ of gypsum

100 ml extract requires;

$$= \frac{(B-S)ml \times 0.01 \times 100}{1000 \times 5} \times 86\, g \text{ of gypsum}$$

= (B – S) x 0.0172 g gypsum

Now; (B – S) x 0.0172 g gypsum is required for 4 g soil

So; 1 kg soil requires → (B – S) 0.0172 x 1000/4 g of gypsum

Then; 2.25 x 10^6 kg soil requires →

$$= \frac{2.25 \times 10^6 \quad \times (B\text{-}S) \times 0.0172 \times 1000}{1 \qquad\qquad\qquad 4}$$

= 9.675 x (*B* - *S*) ton of gypsum

Result: The given alkali soil require --------------- t ha^{-1} of gypsum for its reclamation.

3.21 Exchangeable Acidity

3.21.1 Exchangeable H^+

Exchange acidity includes exchangeable H^+ and Al^{+++} in soil. These are electrostatically bound on the soil colloidal surface and can be replaced by adding another cation to the soil solution. Exchangeable H^+ and Al^{+++} are called the permanent charge acidity or salt replaceable acidity. When replaced to the soil solution, H^+ directly contributes to soil acidity, whereas Al^{++} indirectly contributes to soil acidity by inactivating the OH^- as aluminium monohydroxy [$Al(OH)^{++}$], aluminium dihydroxy [$Al(OH)_2$]$^+$, aluminium hydroxide [$Al(OH)_3$] and tetrahydroxy aluminate [Al] species, depending on the soil pH.

Principles

Hydrogen (H^+) and aluminium (Al^{+++}) on the exchange surface are replaced by adding 1 M KCl to the soil. K^+ replaces the H^+ and Al^{+++} from the exchange surface and itself get adsorbed. Concentration of K^+ in 1 N KCl is sufficiently high to increase the rate of direct reaction much higher than the rate of the reverse reaction for replacement of all the cations including H^+ and Al^{++} from the exchange surface to the soil solution.

The $H^+ + Al^{+++}$ in the KCl extract of soil are then determined by titrating against NaOH in presence of phenolphthalein indicator. At the end of the titration all the H^+ are neutralized and Al^{+++} is neutralized to $Al(OH)_3$ The OH^- bound by Al^{+++} are then released by adding NaF. The F^- form complex with Al^{+++} to release the OH^-. The amount of OH^- released is then determined by titrating against HCl. The meq of H^+ consumed in the titration is equal to the meq of OH^- which in turn equal to the meq of Al^{+++} in the extract. After determining the meq of H^+ + Al^{+++} and only Al^{+++} in the soil extract, the amount of exchangeable H^+ and exchangeable Al^{+++} in the soil (in meq / 100 g) is calculated.

Reagents

1. Potassium chloride (KCl) 1N: Dissolve 74.56 g of potassium chaloride in distilled water and dilute to 1 litre.
2. 0.02 N standard HCl
3. 0.02 N standard NaOH
4. 1N NaF solution : Dissolve 42 g of NaF in distilled water and dilute to 1 litre.
5. Phenolphthalein indicator. Dissolve 1 g of the dry indicator powder in 100 mL of 95% ethanol.

Apparatus required

1. Analytical balance
2. 100 mL conical flask
3. 25 mL volumetric pipette
4. Reciprocating shaker
5. Burette
6. Filter paper

Procedure

Take 5 gram of soil in a 100 mL conical flask. Add 25 mL of 1 N KCl to the soil and shake the contents for 30 minutes by a reciprocating shaker. Then filter the suspension to another 100 mL conical flask using an ordinary (Whatman No.1) filter paper. Add another 25 mL of 1 N KCl to the soil on the filter paper and collect all the filtrate. Then add 3 to 4 drops of phenolphthalein indicator to the KCl extract of soil and titrate against 0.02 N NaOH till a pink colour end point is obtained. The amount of exchangeable H^+ and Al^{+++} is determined from this titration value.

3.21.2 Determination of exchangeable Al^{+++}

Add one to two drops of 0.02 N HCl to make the solution colourless. Then add 10 mL of 1 N NaF. Reappearance of pink colour indicates the presence of Al^{+++}. Titrate against 0.02 N HCl till a colourless end point is obtained. Wait for some times. If pink colour reappears, titrate against 0.02 N HCl till the pink colour disappears and a colourless end point is obtained. Note the total volume of 0.02 N HCl consumed. From this titration value the amount of exchangeable Al^{+++} is determined. Run a blank side by side following all the above steps except taking the soil.

Reactions

$$cly\left|\begin{matrix} H^+ \\ Al^{+++} \end{matrix}\right. + 4KCl \rightleftharpoons Clay\left| 4K^+ \right. + HCl + AlCl_3$$

$$HCl + NaOH \rightleftharpoons NaCl + H_2O \qquad (1)$$

$$AlCl_3 + 3\ NaOH \rightleftharpoons 3NaCl + Al(OH)_3 \qquad (2)$$

$$HCl + AlCl_3 + 4\ NaOH \rightleftharpoons 4NaCl + Al(OH)_3 + H_2O \text{ (Adding eqn. 1 and 2)}$$

$Al(OH)_3 + 6NaF \rightleftharpoons Na_3AlF_6 + 3\ NaOH$
(sodium hexafluroaluminate)

$3\ NaOH + 3HCl \rightleftharpoons 3NaCl + 3H_2O$

Reaction of the phenolphthalein indicator

Phenolphthalein is an acid base indictor which can be represented as HIn (see chapter 11) where In is the organic anionic part of the indicator to which H is attached.

$HIn \rightleftharpoons H^+ + In^-$

(Col. less) (Pink)

The indicator is colourless below pH 8.5 and pink at and above pH 8.5. When the indicator is added to the KCl extract of soil it remains colourless because of the acidic pH due to the presence of H^+ in the extract. When titration is carried out with NaOH, the H^+ in the soil extract is neutralized to H_2O and Al^{+++} is neutralized to $Al(OH)_3$. After neutralization of all H^+ and Al^{+++}, the NaOH added from the burette reacts with the indicator, neutralizing the H^+ at the right side of the equilibrium. As a result, the rate of reverse reaction decreases and the equilibrium shifts from left to right and more of the In^- species are formed to give the pink colour. It indicates that the titration is over.

Before determination of exchangeable Al^{+++}, the indicator is made colourless by adding one or two drops of 0.02 N HCl. With addition of H^+, the rate of reverse reaction increases and the equilibrium shifts from right to left to give the colourless form of the indicator. When NaF is added it reacts with $Al(OH)_3$ with formation of NaOH (see the reactions) and the pH rises above 8.5. The equilibrium shifts to right i.e. the concentration of In^- becomes higher than the concentration of Hin. The indicator, therefore, exhibits pink colour. This is why the appearance of pink colour after adding NaF indicates the presence of Al^{+++} in the extract. If no exchangeable Al^{+++} is present in soil then no $Al(OH)_3$ is formed and no NaOH is formed after adding NaF. The phenolphthalein indictor, therefore, remains colourless indicating the absence of Al^{+++} in the soil extract.

Phenolphthalein is used as the indicator in this experiment because phenolphthalein changes its colourless form to pink at pH 8.5. At this pH Al^{+++} is completely hydrolysed to $Al(OH)_3$. Methyl red exhibits colour change from red to yellow at pH 6.3. At this pH hydrolysis of Al^{+++} is not completed. Most of the Al^{+++} exist in the form of $Al\ (OH)^+_2$ species. Therefore with addition of NaF, the meq of OH^- released is less than the meq of Al^{+++} which creates a negative error. Phenolphthalein is, therefore, used as the indicator in this experiment but not methyl red.

Calculation

1. Exchangeable H^{+}+ Al^{+++} (meq/100 g) $= (S_{NaOH} - B_{NaOH}) \times N_{NaOH} \times \frac{100}{w}$

2. Exchangeable Al^{+++} (meq/100 g) $= S_{HCl} \times N_{HCl} \times \frac{100}{w}$

3. Exchangeable H^{+} (meq/100 g) = Exchangeable (H^{+} + Al^{+++}) (meq/100 g) – Exchangeable Al^{+++} (meq/100 g)

Where,

S_{NaOH} = Sample titration value of NaOH (mL)

B_{NaOH} = Blank titration value of NaOH (mL)

S_{HCl} = Sample titration value of HCl (mL)

N_{NaOH} = Normality of NaOH

N_{HCl} = Normality of HCl

w = Weight of the soil taken (g)

Derivation of the formula for calculating exchange acidity and exchangeable Al^{+++}

A. Exchange Acidity

$S_{NaOH} \times N_{NaOH}$ = meq of NaOH consumed in the sample titration

= mEq of exchange acidity neutralized in the sample

= mEq of exchange acidity in the sample.

$B_{NaOH} \times N_{NaOH}$ = mEq of NaOH consumed in the blank titration

= mEq of exchange acidity in the blank

(The blank titration is carried out to detect H^{+} + Al^{+++}, if any, in the reagents used for this experiment and to rectify accordingly).

meq of exchange acidity extracted from the soil

$= S_{NaOH}\ N_{NaOH} - B_{NaOH}\ N_{NaOH}$

$= (S_{NaOH} - B_{NaOH}) \times N_{NaOH}$

This much mEq of exchange acidity is present in w g of soil taken.

∴ mEq of exchange acidity / 100 g soil

$$= (S_{NaOH} - B_{NaOH}) \times N_{NaOH} \times \frac{100}{w}$$

B. *Exchangeable Al^{+++}*

$S_{HCl} \times N_{HCl}$ = meq of HCl consumed

= meq of NaOH neutralized

= meq of $Al(OH)_3$ forming complex with NaF

= meq of Al^{+++} in the sample

= meq of Al^{+++} in w g of soil

$$\therefore \text{Exch. } Al^{+++} \text{ (meq/100g soil)} = S_{HCl} \times N_{HCl} \times \frac{100}{w}$$

3.22 Phosphate Fixing Capacity of Soil

The P fixing capacity of a soil is an indicative of the reserve soil P. Under suitable systems of soil management fixed P becomes the potential source of soil P and became the available soil P to growing plants. P use efficiency is generally very low in all soils due to the conversion of part of the applied P to unavailable forms through fixation. The extent of P fixation in soils rests on the clay fraction. The formation of Fe and Al phosphates in soils occurs due to the formation of oxides and hydroxides of these metals with P in solution. The soils that dominated with sesquioxide and kaolinite type of clay have a tendency to fix and immobilize larger amount of P from the soluble fraction of soil P. For determination of P fixing capacity of soil the method given by Waugh and Fitts (1966) is widely accepted.

In acid soils:

$$Al^{3+} + H_2PO^-_4 + 2H_2 \rightarrow 2H^+ + Al\ (OH)_2H_2PO_4 \text{ (in soluble)}$$

In neutral and alkaline soils:

$$Ca\ (H_2PO_4)_2 + 2\ CaCO_3 \rightarrow Ca_3\ (PO_4)_2 \text{ (Insoluble)} + 2CO_2 + 2H_2O$$

Principle

A given amount of soil, in series is incubated with varied concentrations of P in solution for a given period of time. The soil is then extracted and P is determined with Olsen and Bray I method as per the reaction of the soil. The recovery of P is subtracted from the added P and the retained or fixed P is calculated as percent of the added quantity.

Instruments and apparatus

1. Spectrophotometer
2. Mechanical shaker
3. Conical flask
4. Volumetric flask
5. Pipette, etc

Reagents

1. *0.5 M $NaHCO_3$* (Olsen's reagent) or Bray extractant No 1 (0.03N NH_4F in 0.025N HCl)
2. *Molybdate reagent* (1.5 % Dickman and Bray's reagent): Dissolve 15 g $(NH_4)_2MoO_4$ in 300 ml of warm distilled water. Add the solution to exact 350 ml of 10 N HCl solution gradually with stirring. Dilute to 1 litre with distilled water.
3. *Activated Charcoal*: Darco G - 60.
4. *Stannous chloride stock solution*: Dissolve 10 g $SnCl_2.2H_2O$ in 25 ml of concentrated HCl. Add a piece of pure metallic tin (Zn metal to prevent oxidation) and store the solution in a glass stoppered amber colour bottle.
5. *Stannous chloride working solution*: Dilute 0.5 ml of the stock solution of stannous chloride to 66.0 ml with distilled water just before use.

Primary P standard of 2500 ppm (stock solution)

Dissolve 9.44 g of pure anhydrous monocalcium phosphate $Ca(H_2PO_4)_2$, in distilled water and make the volume to 1 litre. This is a 2500 ppm P stock solution.

Secondary P standard

Dilute 0, 1, 2, 3, 4, 5, 6, 8. 10 and 15 mL portions of the stock P solution to 50 mL in volumetric flasks to get the solutions containing 0, 50, 100, 150, 200, 250, 300, 400, 500 and 750 ppm of P.

Procedure

1. Weigh 2 g of air dry soil in duplicate in a series of 18 conical flasks of 100 mL capacity.
2. Add 1 mL of each of the 9 dilute standard P solutions to the soil taken in different pairs of the flasks. This will result into concentrations of 0, 25, 50, 75, 100, 125, 150, 200, 250 and 375 µg of added P per g of soil.

3. Plug the conical flasks with cotton wool and allow to incubate at room temperature for 96 hours.
4. Extract the soil P by Olsen's reagent or Bray's reagent as described under available P determinations.
5. Take 5 mL of the clear and colourless filtrate and determine the P content as per the procedures described in available P determination.
6. Work out the extracted amount of P as part of added quantity. Remaining part is the fixed P.
7. Plot the amounts of fixed P against the added quantities of P and locate the point of inflection on the curve.

Calculation

a. Available P content of soil without P added (Say) = $A\ \mu g\ g^{-1}$
b. Available P content of soil with P added (Say) = $B\ \mu g\ g^{-1}$
c. Amount of P extracted out of added P (Say) = $(B - A)\ \mu g\ g^{-1}$
d. Amount of P added (Say) = $C\ ug\ g^{-1}$
e. Amount of P fixed by soil (Say) = $C - (B - A)\ \mu g\ g^{-1}$

$$\text{P fixation (percent of added P)} = \frac{C-(B-A)}{C} \times 100$$

3.23 Potassium Fixing Capacity of Soil

When soluble form of K (K fertilizers) is added to soil, it readily goes into the non exchangeable or the fixed or less and difficulty available form and has restricted the K availability to crops. This conversion is of great significance as it is not a net loss of K but a temporary storehouse of K which becomes available later. This process also prevents the loss of K by leaching to a great extent. Fixation of K in soil depends on soil type, dominating clay minerals, wetting and drying conditions of soil, presence of other cations, available and added K contents, pH, submerged or arable condition, soil texture and soil moisture content.

Instruments Required

Flame Photometer

Reagents Required

1. *Neutral 1 N Ammonium Acetate*: As described in available K estimation
2. *Standard stock solution (1000 mg K/L)*: As described in available K estimation

3. *Standard Solution A (10 mg K/L):* Dilute 1 mL of stock solution to 100 mL
4. *Standard Solution B (100 mg K/L):* Dilute 10 mL of stock solution to 100 mL
5. *Standard series of K solution:* Prepare 0, 5, 10, 15, 20, 25, 30 and 40 mg K L^{-1} solution (100mL each) for standard curve.

Procedure

A) For Coarse textured soil

i). Rice growing soils

1. Weigh 5 g of air dry soil samples in 9 pairs of 100 mL capacity of conical flasks.
2. Add 0, 2.5, and 5.0 mL of solution A and 1.0, 1.5, 2.0, 2.5, 3.0 and 5.0 mL of solution B (each in duplicates)
3. Add measured quantity of water so that each flask contains 10 mL of liquid (10.0, 7.5, 5.0, 8.5, 8.0, 7.5, 7.0 and 5.0 mL of water respectively.
4. Stopper the flasks with cotton wool and allow to stand for 72 hours at room temperature.
5. Add 25 mL of neutral 1N ammonium acetate solution and shake for 5 minutes.
6. Filter through Whatman No. 1 filter paper and determine K content in the clear filtrate, after drawing a standard curve as in the case of available K determination, using flame photometer.

ii). Wheat growing soils

1. Weigh 5 g of air dry soil samples in 9 pairs of 100 mL capacity of porcelain dishes.
2. Add exactly half quantities of K used for rice growing soils through solutions A and B and distilled water.
3. Mix thoroughly by shaking a little and allow the contents to dry in shade for 24 hours.
4. Add 50 % of the rest (i.e., 1/4 th of the total quantity of K and water) and allow to dry for 24 hours.
5. Finally add the remaining part of K (i.e., 1/4 th of the total quantity of K and water) before allowing to dry for another 24 hours.

6. Add 25 mL of neutral 1N ammonium acetate solution and shake for 5 minutes.
7. Filter through Whatman No. 1 filter paper and determine K content in the clear filtrate, after drawing a standard curve as in the case of available K determination, using flame photometer.

B) For fine textured soils

1. Weigh 5 g of air dry soil samples in 11 pairs of 100 mL capacity of conical flasks.
2. Add 0, 2.5, and 5.0 mL of solution A and 1.0, 1.5, 2.0, 2.5and 5.0 mL of solution B and 1.0, 1.5, and 2.0 mL of stock solution (1000 mg K/L) each in duplicates.
3. Add 10.0, 7.5, 5.0, 9.0, 8.5, 8.0, 7.5, 5.0, 9.0, 8.5, and 8.0 mL of distilled water respectively so as to maintain the quantity of liquid at 10 mL in each flask.
4. Plug the conical flasks with cotton wool and keep for equilibration for 72 hours at room temperature.
5. Add 25 mL of neutral 1N ammonium acetate solution and shake for 5 minutes.
6. Filter through Whatman No. 1 filter paper and determine K content in the clear filtrate, after drawing a standard curve as in the case of available K determination, using flame photometer.
7. Plot the extracted K against added K and locate the point of inflection for calculation of K fixed.

Calculation

Amount of ammonium acetate extractable K in untreated soil = A

Amount of ammonium acetate extractable K in treated soil = B

Amount of recovered K (out of the added K) = B-A

Amount of K added = C

Amount of K fixed = C-(B-A)

$$K\ fixation(\%) = \frac{C-(B-A)}{C} \times 100$$

Notes

In case the inflection point on the curve is not sharp enough or is absent even if the procedure is repeated, the mean of the amounts fixed from at least 3 or 4 graded doses may be worked out and used for calculating the per cent K fixed.

4

Basic Concepts of Water Quality

The influence of water quality on the soil and plant growth is related to the chemical and physical properties of the soil, the salt tolerance of the crop, the climatic regime of the area and the method, frequency and the amount of irrigation water applied. Irrigation water quality refers to its suitability and sustainability for use in agricultural crops. The problems that result from the use of a particular irrigation water, vary both in kind and degree, and are modified by soil, climate and crop as well as by the skill and knowledge of the water user. As a result there is no set limit on water quality, rather its suitability for use is determined by the conditions of use which affects the accumulation of water constituents and which may restrict crop yield. The soil problems most commonly encountered and used as a basis to evaluate water quality are those related to salinity, water infiltration rate, toxicity and a group of other miscellaneous problems. Numerous guidelines are available for evaluation of water quality. Time to time some of these guidelines has been updated with modification of some criteria like SAR to Adj. SAR and the Adj. R_{Na} and development of new concepts like water quality Index (WQI) which have increased our predictability. The new guidelines are relied heavily on old ones. The modified guideline by Ayers and Westcot (1985) was found to be the most reliable to predict the water quality for irrigation.

Irrigated agriculture is playing a major role in enhancing food and livelihood security of a country. Supply of fresh water which is an important input for the sustainable and economic development of agriculture are not enough to meet the requirement of all sectors of economy. Reduced water availability to agriculture from the present share of 85% of country's water resources to about 70-75% by 2020AD would affect the capacity of the country to meet the food production target, unless supplies are augmented from unconventional sources. For future agriculture there are no alternatives but to rely on non conventional sources for partial alleviation of the forecasted water scarcity. Among these sources extraction of marginal quality ground water and the waste water generated from urban and industrial activities have the potential to augment the water supplies. Many associated soil, plant and environmental problems come up when these sources of water are directly used for irrigation. For addressing these problems and to get more benefit from these sources of water, the first step is to understand how an irrigation water source affects the soil plant system. Therefore knowledge of irrigation water quality is critical to develop appropriate site and crop specific comprehensive short term and long term management programs.

4.1 Suitability of Irrigation Water

The suitability of irrigation water depends upon several factors such as water quality, soil type, plant characteristics, irrigation method, drainage, climate and local conditions. The integrated effect of these factors on the suitability of irrigation water (SI) can be expressed by the relationship given below;

$SI = QSPCD$

Where;

Q = quality of irrigation water, that is total salt concentration, relative proportion of cations

S = Soil type, texture, structure, permeability, fertility, calcium carbonate content, type of clay minerals, initial level of salinity and alkalinity before irrigation

P = Salt tolerance characteristics of the crop, its variety and growth stage

C = Climate that is, total rainfall, its distribution, and evaporation characteristics

D = Drainage conditions, depth of water table, nature of soil profile, presence of hard pan or lime concretion and management practices

These factors act interactively such that a single suitable criterion is hard to be adopted for widely varying conditions. However Bureau of Indian Standards (BIS) has developed some general broad guidelines (IS:11624-1986) for assessment of water quality which were reaffirmed in 2001 and 2009.

4.2 Water Quality Evaluation

Conceptually water quality refers to its suitability for a specific use and the criteria for judging water quality for different purposes are different. Specific uses have different quality needs and among all sectors of water use, agriculture is most sensitive to water scarcity and water quality.

The objective of water quality evaluation is to assess its suitability for irrigation and assist to cope with potential water quality problems that might reduce production through management. Basing on the experiences and measured responses certain parameters have emerged as indicators of irrigation water quality. These indicators are then organized into guidelines related to suitability for use. Numerous such guidelines are available covering many types of use.

Figure 4.1 is a flow diagram suggesting a concept for assessing water quality. Plant tolerance allows a wide range of options as to suitable crops and this feeds back into management aspects which determine whether these crops are profitable or whether the yield losses due to salinity or toxicity are acceptable. Under soil factors where quality is marginal, there is a possibility of adding ameliorants to the soil or changing to another soil type to avoid problem.

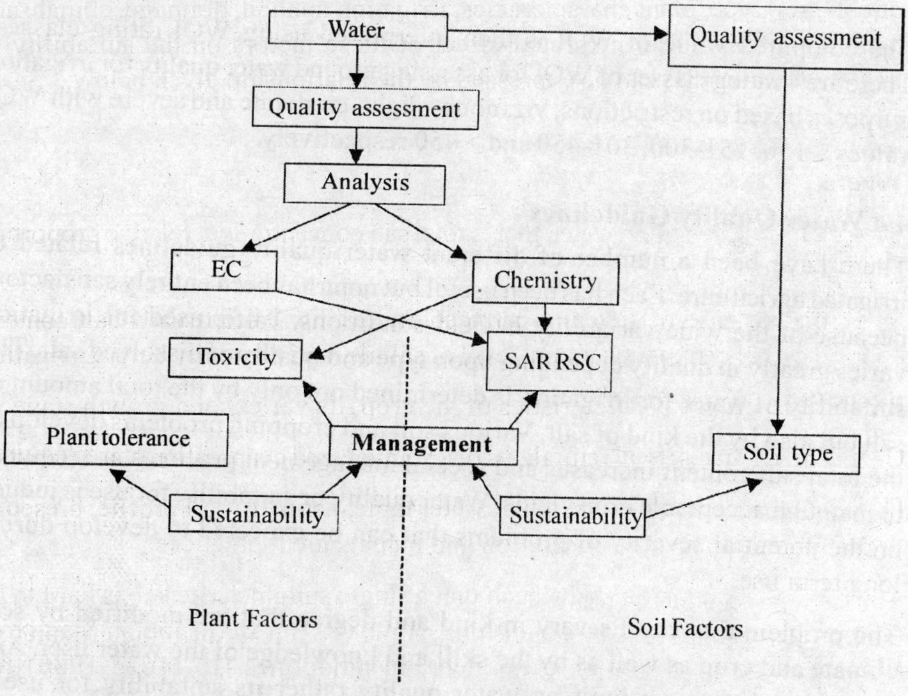

Fig. 4.1: Flow diagram of evaluation of water quality

4.3 Water Quality Index (WQI)

Water quality index is a new tool for quantitative assessment of water quality in relation to a particular function. It is computed as;

$$WQI = \sum_{i=1}^{n} W_i \times q_i$$

Where; W_i = Relative weight assigned to a particular quality parameter calculated on the basis of its relative importance as computed to other parameter.

$$Wi \frac{w_i}{\sum_{i=1}^{n} w_i}$$

Where, w_i= the weight of each parameter and n is the number of parameters

q_i = the quality rating scale of each parameter is computed by dividing the observed value by its standard value as per the guidelines laid down in the BIS 10500, 1991 or FAO and multiplying with 100.

$$q_i = \frac{C_i}{S_i} \times 100$$

The computed value of WQI is then interpreted using WQI rating classes. There are 4 rating classes of WQI for assessing ground water quality for irrigation purposes based on restrictions, viz, none, slight, moderate and severe with WQI values < 150, 151-300, 301-450 and > 450 respectively.

4.4 Water Quality Guidelines

There have been a number of different water quality guidelines related to irrigated agriculture. Each has been useful but none has been entirely satisfactory because of the wide variability in field conditions. Water used for irrigation varies greatly in quality depending upon type and quality of dissolved salt. The suitability of water for irrigation is determined not only by the total amount of salt but also by the kind of salt. Various soil and cropping problems develop as the total salt content increases and special management practices are required to maintain acceptable crop yields. Water quality or suitability for use is judged on the potential severity of problems that can be expected to develop during long term use.

The problem that results vary in kind and degree that are modified by soil, climate and crop as well as by the skill and knowledge of the water user. As a result there is no set limit on water quality rather its suitability for use is determined by the conditions of use which affect accumulation of the water constituents and which may restrict crop yield. The soil problems that are most commonly encountered and used as a basis to evaluate water quality are those related to salinity, sodicity, toxicity of ions and a group of other miscellaneous problems.

(i) ***Salinity***: It is related mainly to osmotic potential and its effect on crop growth. It is determined by salt concentration rather than specific constituents. The guideline indicates that it is necessary to consult detailed tolerance tables to determine plants which would not suffer unacceptable yield loss.

(ii) ***Sodicity***: The sodic (sodium) hazard is related to the detrimental effect of the exchangeable sodium percent (ESP) on soil structure and the direct toxic effect on sodium sensitive plants.

(iii) *Toxicity*: This is the specification effects of solutes (excluding sodium) of a nutritional nature, e.g. boron, chlorine and certain metals.

4.5 Guidelines for Irrigation with Waste Water

Special consideration for water quality guidelines as suggested are as follows.

1. Use gypsum when saline waters (having SAR > 20 and Mg/Ca ratio > 3 and rich in silica) induce water stagnation during rainy season and crops grown are sensitive to it.
2. Leaving the field fallow during the rainy season is helpful when SAR > 20 and waters of higher salinity are used in low rainfall areas.
3. Additional phosphorus fertilization is beneficial, especially when / ratio in water is > 2.0
4. Canal water preferably be used at early growth stages including pre-sowing irrigation for conjunctive use with saline waters.
5. If saline water is to be used for seeding of crops, 20% extra seed rate and a quick post-sowing irrigation (within 2-3 days) will ensure better germination.
6. When $EC_{iw} < EC_e$ (0 - 45 cm soil at harvest of rabi crops), saline water irrigation just before the onset of monsoon will lower soil salinity and will raise the antecedent soil moisture for greater salt removal by rains.
7. Use of organic materials in saline environment enhances yields.
8. Accumulation of B, NO_3, Fe, Si, F, Se and heavy metals beyond critical limits proves toxic. Expert advice prior to the use of such waters may be obtained.
9. For soils having (i) shallow water table within 1.5 mm in kharif season and (ii) hard sub soil layers, the next lower EC (alternate mode of irrigation canal/saline) is applicable.

Numerous guide lines are available for judging irrigation water quality. Each has been useful but none has been entirely satisfactory because of the wide variability in field condition. Time to time some new guidelines have been developed with modification of the previous ones to give more practical procedures for evaluating and managing water quality related problems of irrigated agriculture. Soil problems such as salinity, sodicity, toxicity of boron and chloride still form the basis for evaluation of irrigation water quality. However, there is variation with respect to their measurement and way of interpretation either by taking a single factor or a number of factors together in relation to soil and climatic variability. The prevailing guidelines need to be updated depending on the use and suitable situation specific management practices need to be developed to address the soil and cropping problems on regional basis.

4.6 Collection, Processing and Storage of Water Sample

Theory

Liquid waste flowing out of a factory, farm, commercial establishment or household into a water body such as a river, lake or lagoon or a sewage system or reservoir is called effluent. Different waste materials like smoke, liquid industrial refuse or sewage discharged into the environment also included under pollutants can be used to determine the degree of pollution. Improper sampling does not represent the irrigation or waste water being sampled. Proper sampling followed by accurate chemical, physical and biological analysis of irrigation/ effluent samples helps in generation of authentic data for meaningful interpretation of results.

Requisites

1. *Sampling bottle:* 500 mL or 1 L new PVC bottles to be used except for bacteriological, oil based or solvent analysis.
2. Sampling tools:
   (i) Sampling hand pump with extension tube for depth sampling at low flow
   (ii) Sampling beaker (250 Ml or 500 mL or 1L) with screw in extension rods to be used for depth sampling with sufficient flow.
3. Gloves
4. Nylon nets
5. Marker pen

Collection of water/effluent sample

Water/effluent samples are collected by two ways to characterize the chemical and biochemical parameters; Grab Sample and b) Composite sample

Grab Samples

A grab sample is one that is taken to represent the performance or characteristics at one moment in time when the sample is collected and is not mixed with any other samples. It is an individual discrete sample collected manually by removing a quantity of water/effluent from a flow stream at a single point in time during the flow day. As the name implies, "grab samples" are simple scoops of the water being sampled and are appropriate where water samples are well mixed and conditions are constant or slow to change. This sample is used for instant measurement of water/sludge. It is typically used for analysis of unstable parameters such as temperature, dissolved oxygen or pH.

Composite Samples

Composite samples are either amalgamated or made up of smaller sub samples. It shows the average composition of a series of grab samples. It is collected in two ways;

a) ***Time weighted***: When a composite sample is prepared by combining a series of individual discrete of a flow stream over a set time is called time weighted composite sample. Sub samples of equal volume taken at specific time intervals i.e., every hour composited to make a single daily sample. A composite sample representing a 24 hours period is considered standard for most determinations.

b) ***Flow weighted***: When a composite sample is prepared by combining a series of individual discrete proportional to flow is called flow weighted composite sample. A properly designed sampling port is needed to ensure that the sample is representative of the effluent flow stream. It is proportional to the flow and is usually linked to an automatic flow meter. It is extremely accurate and can be used to establish the total water/ waste water load. This method of sampling is preferable.

Sampling Procedure

1. Clean and sterile polyethylene bottles of 500 mL to 1000 mL capacity are used to collect the effluent samples from effluent/irrigation channels/pipes.
2. Samples are collected in duplicates to avoid error in estimation.
3. It is collected on monthly basis for any routine analysis from the discharge point, 500 m upstream and 500 m downstream of the discharge point.
4. Samples after collection is transported in a box containing ice packs to the laboratory for analysis.
5. To neutralize the effect of chlorine on the microflora 0.1 % sodium thiosulphate is added to the final composite sample bottles.
6. All sampling containers should be clearly labeled to include the following information;

a. A sample identification number

b. The source and location of sample collection

c. The date and time of sample collection

d. The name of the sample collector/ representative

e. Parameters to be analysed

f. Signature of the sample collector/ representative

Sample Processing and Storage

a. After collection of samples, it is processed within 6 hours of sample collection in the laboratory.

b. One litre samples in duplicates are to be filtered in the laboratory through 180, 60 and 20 µm pore size nylon nets to separate the suspended organic loadings. In case of less suspended materials filter papers may be used to filter it.

c. The 20 µm pore size nylon nets are used to collect the samples for microbiological analysis.

d. Processed water/ effluent samples are preserved at a low temperature (6-10°C) and analyzed in accordance with the guidelines provided by the laboratory.

Physicochemical Analysis

All the equipments used in analysis are to be checked and appropriately calibrated according to the manufacturer's instructions. pH, EC, total dissolved solids (TDS), total suspended solids (TSS), dissolved oxygen (DO), biochemical oxygen demand(BOD), chemical oxygen demand (COD) are to determined. Different dissolved ions like Ca &Mg, Na, K, Cl, carbonates and bicarbonates, nitrate, sulphates etc are to be analyzed. Microbial groups like bacteria and fungi populations along with other disease causing organisms can be determined.

Precautions

1. Recent rain can be avoided as it dilutes irrigation water/waste water.
2. Absence of wind flow that may drive some constituents towards one side of the water bodies may be avoided.
3. Shading from cloud and vegetation can influence the level of dissolved oxygen.

4. Personal safety; use of disposable gloves, eye protection, protective clothes and used of sanitizer for hand cleaning etc.

4.7 Total Suspended Solids (TSS) in Water Sample

Theory

The term total suspended solid can be referred to materials which are not dissolved in water and are not filterable in nature. It is defined as residue on a filter paper upon evaporation of non filterable parts of a sample. It can be settled out of the water column onto the bottom of water stream when stream velocities are low. It includes silt, clay, plankton, organic waste, inorganic precipitates, etc. The quantity of suspended solid is a measure the quality of irrigation water/waste water. The residue on filter paper is evaporated to dryness in a pre weighed dish and dried to constant dry weight at 105 °C. The increase in filter paper weight represents the total suspended solids.

Materials Required

1. Electronic balance
2. Hot air oven
3. Aluminium weighing dish
4. Filter paper
5. Forceps/Tweezers
6. Sample collection container
7. Measuring cylinder
8. Desiccator

Procedure

1. Measure exactly 100 mL of water/effluent sample from the sample container and filter it through an oven dried pre weighed filter paper. Record the oven dried weight of the filter paper.
2. After filtration, remove the filter paper with residues and transfer it into an aluminium dish for drying.
3. Keep the aluminium dish with filter paper in hot air oven at 105 °C to completely evaporate the moisture.
4. After evaporation of moisture remove the dish from the oven and cool it in desiccators. Take the weight of the oven dried filter paper with residues.

5. Repeat the cycle of drying, cooling and weighing until a constant weight of the oven dried filter paper with residues is achieved.
6. Record the final weight in a standard tabular form to calculate the total suspended solids content in water/effluent sample.

Calculation

$$\text{Total suspended solids (TSS) in mg/L} = \frac{W_2 - W_1}{V} \times 1000$$

Where;

W_1 = Weight of oven dried filter paper (mg)

W_2 = Weight of oven dried filter paper + residues (mg)

V = Volume of sample taken (mL)

Conclusion: The total suspended solids (TSS) is calculated and expressed in mg/L.

Observation Table

Sl No	Source of water/ effluent	V (mL)	W_1 (mg)	W_2 (mg)	TSS (mg/L)
1					
2					
3					

4.8 Total Dissolved Solids (TDS) in Water Sample

Theory

The total dissolved solid refer to the material that are completely dissolved in water. These solids are filterable. Hence dissolved solids are those that can pass through a filter paper. It includes some organic materials as well as salts, inorganic nutrients and toxins. The filtrate is evaporated to dryness in a pre weighed dish and dried to constant dry weight at 105°C. The increase in dish weight represents the total dissolved solids.

As per United States Environment Protection Agency (USEPA) a limit of 500 mg dissolved solids/L is set as secondary standard for drinking water. The total dissolved solids (TDS) of different water bodies are;

Water bodies	TDS (mg/L)
Fresh water	< 1500 (100-1500)
Blackish water	1500 to 5,000
Saline water	> 5,000
Waste sludge	2000 to 12,000
Ocean	35,000

Environmental significance

Dissolved material, mineral, gases, organic constituents may produce aesthetically displeasant water, taste and odour. Some dissolve organic chemicals may deplete the dissolved O_2 in the receiving water. Water with higher solid content often has an adverse effect on human being. Sometimes the reverse effect may cause death of lower organisms. This parameter is measure to find out the suitability of water for living world.

To determine the TDS in water, a known volume of filtered water sample is evaporated to dryness in a pre weighed container at 105°C and the increase in weight of the container after evaporation of water represents the TDS content of the sample evaporated.

Materials Required

1. Electronic balance
2. Porcelain evaporating dish with 100 mL capacity
3. Hot air oven/ Muffle furnace
4. Filter paper
5. Forceps/Tweezers
6. Sample collection container
7. Measuring cylinder
8. Desiccator

Procedure

1. After filtration of water sample, exactly 50 mL of filtrate is transfered it into an oven dried pre- weighed porcelain dish.
2. Place the dish with sample in a hot air oven at 105 °C to completely evaporate the moisture or in a muffle furnace to completely dry it.

3. After evaporation of moisture, remove the dish from the oven and cool it in a desiccator. Take the weight of the oven dried dish with residues.
4. Record the final weight in a standard tabular form to calculate the total dissolved solids content.

Calculation

$$\text{Total dissolved solids (TDS) in mg/L} = \frac{W_4 - W_3}{V} \times 1000$$

Where;

W_3 = Weight of oven dried porcelain dish (mg)

W_4 = Weight of oven dried porcelain dish + residues (mg)

V = Volume of filtrate taken (mL)

Conclusion: The dissolved solids (TDS) in water sample is determined and is expressed in mg/L.

Observation Table

Sl No	Source of water/ effluent	V (mL)	W_3 (mg)	W_4 (mg)	TDS (mg/L)	TS* (mg/L) = TSS+ TDS
1						
2						
3						

*TS = Total solids

4.9 pH and Electrical Conductivity (EC) of Water Sample

4.9.1 Hydrogen Ion Activity (pH)

For assessing irrigation water quality very less attention has generally been given to pH. The normal range of pH for irrigation water as per *FAO (1985)* guidelines is 6.5-8.5. The pH values in surface and ground water used for irrigation ranges from 6.71 to 8.07 and 6.20 to 6.71 respectively. The average pH of canal water and river water ranges from 8.1 to 8.3 and 8.4 to 8.9 respectively. Irrigation water with a pH outside the normal range may cause a nutritional imbalance or may contain a toxic ion. The water with high RSC has high pH and land irrigated by such waters becomes infertile owing to deposition of sodium carbonate.

The pH is the negative logarithm of the hydrogen ion (H^+) activity of a solution. It measures the acidity or alkalinity of a solution. It is a simple but very important estimation for soils, water and effluent samples since pH influences to a great extent the availability of nutrients to crops. It also affects microbial population in soils ecosystem.

The pH scale of water is used to measure the pH of the soil solution/effluent samples since water is the solvent in the solution. Water weakly dissociates to H^+ and OH^-.

According to Law of Mass Action;

$$H_2O \rightleftharpoons H^+ + OH^-$$

At equilibrium, $k_f a_{H_2O} = K_r a_{H^+} \times a_{OH^-}$

k_f = Rate constant for the forward reaction

k_r = Rate constant for the reverse reaction

Taking logarithm to At 25°C; $-\log a_{H^+} + (-\log a_{OH^-}) = 14$ [$k_f H_2O = 10^{-14}$]

Hence; pH + pOH = 14

Therefore, an inverse relationship exists between pH and pOH. If the pH of a solution is 5, then pOH of that solution is 9.

Materials required

1. Buffer solutions of pH 4, 7 and 9.2
2. pH meter with a range of 0-14 pH with electrode assembly
3. Pipette/dispenser
4. Beaker
5. Filter paper

Procedure

For determination of pH

1. Standardize the pH meter with known buffer solutions of pH 4.0, 7.0 and 9.2 respectively.
2. Take 50 mL of the water sample in a beaker. Filter the sample if required.
3. Insert the glass electrode assembly of the pH meter into the water sample taken in the beaker.

4. Take the pH meter reading of the water sample.
5. Record the pH meter readings of the sample in a standard tabular form to calculate the pH of the water sample.

4.9.2 Salinity Problems

A salinity problem due to water quality occurs if salts from applied irrigation water accumulate in crop root zone beyond certain limits and yields are affected with shallow water tables. A salinity problem may also exist due to upward movement of salts from ground water as the water evaporates from the soil or used by the crop.

The total concentration of soluble salts in irrigation water can be adequately expressed in term of electrical conductivity (EC). The basic unit for EC is deci Siemen per meter (dS m^{-1}). The electrical conductivity values of irrigation water can also be converted into other units such as;

(i) EC (dS m^{-1}) × 640 = Total salt concentration, mg L^{-1} or ppm

(ii) EC (dS m^{-1}) × 0.36 = Osmotic pressure, bar

(iii) EC (dS m^{-1}) ×10 = Total cation or anion concentration, meq L^{-1}

Irrigation water classification for salinity hazard used as a guidelines as proposed by Richards (1954) is presented in Table 4.1.

Table 4. 1: USDA classification of irrigation water salinity

Salinity class and description	EC range (dSm^{-1})	Equivalent salt concentration (mg L^{-1})
C1 Low salinity water-good for most soils and crops	< 0.25	< 200
C2 Medium salinity water-some leaching for sensitive crops	0.25-0.75	200-500
C3 High salinity water tolerant crops and leaching required	0.75-2.25	500-1500
C4 Very high salinity water only for permeable soils and tolerant crops	> 2.25	> 1500

4.9.2.1 Determination of Electrical Conductivity (EC)

Electrical conductivity (EC) is the quantity of electricity in coulombs that passes per second per cm^3 of a solution under an applied potential difference of 1 volt. Normally it report the specific conductance of the solution and is expressed in millimohs cm^{-1} or dS m^{-1} (deciSiemen m^{-1}). Current flows in the solution/ irrigation water due to presence of soluble salts like Cl^-, and of Na^+, K^+ Ca^{++} and Mg^{++}. These salts fully dissociate in water to their constituent cations and anions and the current is carried by those cations and anions in the solution. Therefore, electrical conductivity of solution/irrigation water increases with increase in soluble salt content in soil/irrigation water. Electrical conductivity of the saturation extract of the saline and saline-alkali soil/solution/irrigation water is > 4 dS m^{-1}. The electrical conductivity of solution also increases with increase in temperature @about 2% per degree rise in temperature. So it is usually reported at 25°C. Specific soil management practices are required for reclamation of saline and saline-alkali soil ecosystem if contaminated with saline irrigation water to obtain higher yield potential.

Reagents Required

1. Electrical conductivity meter
2. Pipette/dispenser
3. Beaker
4. Filter paper
5. Thermometer
6. Standard 0.02 M KCl : Dissolve 1.491 g of Analytical grade KCl (dried at 110°C in oven) in water and dilute to one litre. This solution has a conductivity of 2.768 mmhos cm^{-1} at 25°C

Procedure

1. Before measuring the conductance of the irrigation water, determine the cell constant of the instrument or standardize the conductivity meter by measuring the conductance of a 0.02 M KCl solution.
2. Take 50 mL of the water sample in a beaker. Filter the sample if required.
3. Insert the conductivity cell of the conductivity meter into the effluent sample taken in a beaker.
4. Note down the temperature of the irrigation water by a thermometer.
5. Take the conductivity meter reading of the sample.
6. Record the conductivity meter readings of the water/effluent samples in a standard tabular form to calculate the conductivity of the water sample.

Electrical Conductivity (dS/m) at 25°C = EC at measured temperature x Temperature factor

Conclusion/Results: The electrical conductivity is to be determined and expressed in dS/m.

Table 4.2: Temperature factor (tf) for determining conductivity at 25°C

°C		tf	°C		Tf
15.0	-	1.247	31.0	-	0.890
16.0	-	1.218	32.0	-	0.873
17.0	-	1.189	33.0	-	0.858
18.0	-	1.163	34.0	-	0.843
19.0	-	1.136	35.0	-	0.829
20.0	-	1.112	36.0	-	0.815
21.0	-	1.087	37.0	-	0.801
22.0	-	1.064	38.0	-	0.788
23.0	-	1.043	39.0	-	0.775
24.0	-	1.020	40.0	-	0.763
25.0	-	1.000			
26.0	-	0.979			
27.0	-	0.960			
28.0	-	0.943			
29.0	-	0.925			
30.0	-	0.907			

Observation Table

SL No	Source of water	Volume of water taken (mL)	pH of the water	EC at measured temperature	Temperature factor	EC at 25 °C
1						
2						
3						

Recommendations for suitable crops to be grown

Table 4.3: Promising varieties for alkali tolerance from CSSRI and other places

Crops	pH and ESP	Varieties/ Genotype
Rice	9.4 and 10.29.4 and 9.8	CSR 10CSR 13, CSR 23, CSR 27, CSR 36, CSR 30
Wheat	9.2 and 9.3	KRL 1-9, KRL 19, KRL 99, WH 157, Raj. 3077
Mustard	12 and 38 (ESP)	Pusa bold, Varuna, Kranti, CSR 52, CSR 54, CS 56
Chickpea	Upto 9.3 pH	Karnal Chana 1
Barley	Upto9.3 pH	CSB 1, CSB 2, CSB 3, DL 200, DL 348, Ratan, BH 97, AZAD
Dhaincha		CSD 137, CSD 123

4.10 Dissolved Oxygen (DO) in Water Sample

Theory

Oxygen is required for all metabolic activities of living organisms that helps to generate energy. Aerobic organisms and processes cannot be completed without oxygen. Oxygen is also a limiting factor for aquatic ecosystem. Atmospheric exchange of gaseous oxygen and oxygen produced from the photosynthetic activities of phytoplankton are the main source of oxygen in water bodies. Dissolved oxygen content less than 4 mg/L in aquatic ecosystem affects fish population. With the increase in depth the dissolved oxygen content also decreases in water bodies. Water temperature also affects the dissolved oxygen content and with increase in water temperature dissolved oxygen content decreases. At 0°C the dissolved oxygen content in a water body will be 14 mg/L and it drops to 7 mg/L at 35°C. In pond and river water dissolved oxygen content is more than the sewage and industrial effluent. In aquatic bodies dissolved oxygen is used for the decomposition of suspended organic loadings.

Principle

Dissolved oxygen present in water is estimated by a method known as Winkler's method/ Idometric method/ Oxygen displacement method. This method is based on addition of divalent Mn solution followed by strong alkali. Dissolved oxygen oxidises the divalent Mn to its higher valency which is precipitated as a brown coloured oxides of manganese, MnO_2 after addition of alkali-iodide solution. In presence of iodide solution in acidic medium (after acidification) the oxidized Mn reverts to divalent state with the liberation of iodine which is equivalent to the original DO content. The iodine so released is calculated by titrating against standard solution of sodium thiosulphate using starch as indicator.

Reactions

$MnSO_4 + KOH \rightarrow Mn(OH)_2 + K_2SO_4$

$$Mn(OH)_2 + \frac{1}{2}O_2 \rightarrow MnO_2 + H_2O$$

$MnO_2 + H_2SO_4 + 2Kl \rightarrow MnSo_4 + K_2\ SO_4 + H_2O + I_2$

$I_2 + Na_2S_2O_3 \rightarrow Na_2S_4O_6$ + 2Nal (in presence of starch indicator)

(Sodium thiosulphate) (Sodium tetrathionate)

Requisites

1. BOD Bottle and other glassware
2. Sodium thiosulphate: Prepare 0.01N aqueous solution (Equivalent weight = 248.18 g)
3. Manganous sulphate ($MnSO_4$) solution: 364g of $MnSO_4 . H_2O$ in 1L distilled water.
4. Alkali- iodide-azide solution (500 g NaOH+ 135 g NaI+ 10 g NaN_3 in 1 L distilled water)
5. Starch indicator: Prepare 1 % aqueous solution (Should be prepared freshly).
6. Concentrated Sulphuric acid (H_2SO_4)

Procedure

1. In a 300 mL capacity BOD bottle 50 mL of water sample was taken.
2. To the sample 1 mL of $MnSO_4$ solution was added.
3. To this 1 mL of alkali - iodide- azide solution was added. It was left undisturbed for 1 minute to give brown colour precipitation.
4. To the solution 1 mL of concentrated sulphuric acid (H_2SO_4) was added and mixed thoroughly to dissolve the brown precipitation.
5. It was then titrated against a solution of 0.01 N sodium thiosulphate till a straw or pale yellow colour appears.
6. To this a few drops of starch indicator was added (colour changed to blue) and titration was continued till the colour of the solution becomes colourless from violet blue.
7. Titre values were recorded in a standard tabular form to calculate the DO content in water/effluent sample.

Calculation

$$\text{Dissolved oxygen (DO) content in mg/L} = \frac{T \times N \times E}{V} \times 1000$$

Where;

T = Titre value/ volume of 0.01 N sodium thiosulphate consumed during titration (mL)

N= Normality of sodium thiosulphate used (0.01 N)

E = Equivalent weight of oxygen (8 mg)

V = Volume of sample taken (mL)

Conclusion DO of the water is determined and expressed in mg L^{-1}

Observation Table

SL No	Source of water	Volume of water taken (mL)	Volume of 0.01N $Na_2S_2O_3$ consumed during titration (mL)			Indicator used and colour changed	DO content (mg/L)
			IBR	FBR	Difference		
1							
2							
3							
4							

4.11 Biochemical Oxygen Demand (BOD) in Water Sample

Theory

The biochemical oxygen demand (BOD) of sewage or effluent or water sample indicates the amount of organic compounds present in it. It is considered as a good index of organic pollution. High amount of organic matter in water leads to higher oxygen utilization by bacteria for degrading it. Use of sewage having high BOD increases the concentration of soluble organic compounds in the aquatic body where it is discharged. Digestion of these compounds in natural ecosystems can deplete available oxygen, thus leading to asphyxiation of fish.

Principle

The determination of BOD is based on the measurement of volume of oxygen required by bacteria to metabolize the organic compounds. The BOD is generally measured by incubating the water sample at 20°C for five days in the dark in aerobic conditions. In the tropical and sub-tropical climate, where both temperature and rate of metabolic activities are high, the incubation is preferably done at 27°C for three days. Since a significant amount of oxygen is required for nitrification, there is an over-estimation of BOD, if it is not checked by a suitable method, e.g. by adding 1 ml of 0.5% solution of allyl thiourea.

Instruments

1. pH meter
2. BOD incubator

Reagents required

1. Allyl thiourea: Prepare 0.5% solution.
2. Hydrochloric acid: Prepare approximately 1 *N* solution
3. Sodium hydroxide: Prepare approximately 1 *N* solution
4. Sodium thiosulphate: Prepare 0.01N aqueous solution (Equivalent weight = 248.18 g)
5. Manganous sulphate solution
6. Alkali- iodide-azide solution ($NaOH+ NaI+NaN_3$)
7. Starch indicator: Prepare 1 % aqueous solution (Should be prepared freshly).
8. Concentrated Sulphuric acid

Procedure

1. Neutralize the water/ effluent sample using 1 N acid or 1 N alkali (to attain a pH of 7.0).
2. Fill the water/effluent sample in 6 BOD bottles (250 ml capacity) without bubbling.
3. Add 1 ml of allyl thiourea to each bottle.
4. Add 2 ml of manganous sulphate and alkali- iodide-azide solutions in succession to three of six BOD bottles, right at the bottom with separate pipettes and replace the stopper.
5. Shake the bottles in the upside down direction at least six times.
6. Allow the brown precipitate to settle.
7. Add 2ml of conc. H_2SO_4 and shake the stoppered bottle to dissolve the brown precipitate.
8. Take 50 ml aliquot in a 250 ml conical flask and titrate with sodium thiosulphate solution till the colour changes to pale straw.
9. Add 2 drops of starch solution to the flask. It will change the colour from pale to blue.

10. Titrate again till the blue colour disappears.
11. Calculate the dissolved oxygen in mg L^{-1}, as given below.
12. Take the mean of dissolved oxygen of the three bottle (D_1).
13. Incubate the rest three BOD bottles at 27^{0}C in a BOD incubator for three days.
14. Estimate the oxygen concentration in all the three incubated samples and take mean (D_2).

Calculation

$$\text{Dissolved oxygen (mg L}^{-1}) = \frac{T \times N \times 8}{V} \times 1000$$

Where

V= Volume of sample taken,

T= Total titre value (mL), and

N= Normality of titrant (i.e. 0.01 *N*).

BOD (mg L^{-1}) = $D_1 - D_2$

Where, D_1 and D_2 are dissolved oxygen contents in the initial (un-incubated) and the incubated samples, respectively.

Conclusion: BOD of the water is determined and expressed in mg L^{-1}

Observation Table

Sl No	Source	Volume of water taken (mL)	Volume of 0.01N Sodium thiosulphate consumed (mL)			Indicatorused and colour changed	D_1 (mg/L)	D_2 (mg/L)	BOD (mg/L)
			IBR	FBR	Difference				
1									
2									
3									

4.12 Chemical Oxygen Demand (COD) in Water Sample

Theory

With the alarming increase in pollution through the discharge of large amounts of various chemically oxidizable organic substances entering the aquatic systems, the biochemical oxygen demand alone does not give a precise estimate of the organic load of the water sample. Besides this, the presence of various toxicants in the sample may severely affect the reliability of BOD results. Therefore the chemical oxygen demand (COD) is considered as a better indicator of organic matter.

Principle

The organic matter content in water/effluent is estimated by their oxidability by using chemical oxidants like potassium dichromate. A known volume of $K_2Cr_2O_7$ first used to oxidize the organic matter and then the excess of oxygen is allowed to react with potassium iodide. This liberates iodine in amounts equal to the excess oxygen. The liberated iodine is estimated by titrating against standard solution of sodium thiosulphate using starch as indicator.

Reactions

$K_2Cr_2O_7 + 4H_2SO_4 \rightarrow K_2SO_4 + Cr_2\,(SO_4)_3 + 4H_2O + 3O$

$C + 7O + 2KI + H_2SO_4 \rightarrow CO_2 + K_2SO_4 + H_2O + I_2$

$I_2 + Na_2\,S_2O_3 \rightarrow Na_2S_4O_6 + 2NaI$ (In presence of starch indicator)

Apparatus

Water bath

Reagents Required

1. Potassium dichromate solution (0.1N)
2. Sodium thiosulphate solution (0.1N)
3. Sulphuric acid (2N)
4. Starch solution (1%)
5. KI solution (15%)

Procedure

1. Transfer 50 mL water/effluent sample in triplicate in 250 mL conical flask.
2. Carry out a blank with distilled water (in triplicate).
3. Add 5 mL of $K_2Cr_2O_7$ solution to each flask.
4. Place the flasks in water bath at 100°C for one hour.
5. Cool the flasks for 10 minutes and add 5 mL of KI solution followed by 10 mL of H_2SO_4 to each flask.

6. Titrate the contents against 0.1N sodium thiosulphate solution until pale yellow colour appears.
7. Add 1 mL of freshly prepared starch solution (it gives blue colour).
8. Titrate again until blue colour disappears completely.
9. Find out the total volume of titrant consumed and take the mean of the three flasks of each sample.
10. Find out the total volume of the titrant used for the blank also.

Calculation

$$\text{COD (mgL}^{-1}) = \frac{8 \times C \times (B - A) \times 1000}{V}$$

Where

C = Concentration of the titrant used (0.1N)

A = Titre value of sample

B = Titre value of blank, and

V = Volume of effluent sample taken

Conclusion: COD of the water is determined and expressed in mg L^{-1}

Observation Table

Sl No	Source	Volume of water taken (mL)	Volume of 0.1N Sodium thiosulphate consumed (mL)			Indicatorused and colour changed	COD (mg/L)
			IBR	FBR	Difference		
1							
2							
3							
4							

4.13 Carbonates (CO_3^{2-}) and Bicarbonates (HCO_3^-) in Water Sample

Theory

Alkalinity is important for fish and aquatic life because it protects or covers against rapid pH change. Higher alkalinity level in surface will cover acid rain and prevent pH change that would otherwise harmful to aquatic lives. Large amount of alkalinity imparts bitter taste in water. The principal rejection of alkaline water or effluent is that it causes corrosion in pipes, affects aquatic animals and also human beings.

Principle

Carbonates and bicarbonates in water/effluent sample can be determined by titrating against standard H_2SO_4 (0.1 N). When phenolphthalein indicator is added to the water/effluent sample, pink color appears if carbonates are present. Then the contents are titrated against standard sulphuric acid till color disappears. That means the carbonates are converted to bicarbonates. The standard acid utilized for conversion of CO_3^{2-} to HCO_3^- is equal to the acid needed for conversion of HCO_3^- to CO_2 and H_2O. Hence, the titre value of carbonates will be obtained by multiplying the titre value of phenolphthalein with two. When one or two drops methyl orange indicator is added to the same contents, straw yellow color appears, which then be titrated against standard sulphuric acid till the color changes to rose red. Here, the bicarbonate derived from carbonates and the bicarbonates originally present in the sample are completely neutralised to CO_2 and H_2O. So, the titre value for HCO_3^- is obtained by subtracting the phenolphthalein titre vale from methyl orange titre value.

$2\ Na_2CO_3 + H_2SO_4 \rightarrow 2\ NaHCO_3 + Na_2SO_4$ (Phenolphthalein)

$2\ NaHCO_3 + H_2SO_4 \rightarrow Na_2SO_4 + H_2O + 2CO_2$ (Methyl orange)

Glassware required

Burette, pipette and conical flask

Reagents required

1. *H_2SO_4 (0.1 N)*: Dilute 2.8 mL of conc. Sulphuric acid to 1L and standardize against 0.1 N Na_2CO_3
2. *Phenolphthalein*: Dissolve 0.25 g of indicator powder in 100 mL of 60 % ethanol
3. *Methyl orange indicator*: Dissolve 0.5 g of indicator powder in 100 mL of 95% ethanol

Procedure

1. Pipette out 10 ml of water/effluent sample into a conical flask.
2. Add 1 or 2 drops of phenolphthalein indicator; the solution will turn pink.

3. Titrate the contents against the 0.1 N H_2SO_4 mounted in burette; the end point is indicated by the disappearance of pink colour. Record the burette leading.

4. If on addition of phenolphthalein the sample does not turn pink, it indicates the absence of $CO3^{2-}$. In such case add methyl orange indicator to the same aliquot and titrate against H_2SO_4 till the straw yellow colour changes to rose red.

5. To the same contents, add 2-3 drops of methyl arrange indicator; solution turns straw yellow colour.

6. Again titrate the contents against H_2SO_4 mounted in burette. The end point of titration is indicated by the development of rose red colour. Note down the reading of burette.

7. Repeat the titrations till two consecutive and concurrent titre values are obtained.

Calculations

$$\text{Carbonate content (meq/L)} = 2y \times N \times \frac{1000}{\text{mL of sample taken (mL)}}$$

Where;

y = Titre value of the sample for indicator phenolphthalein

N = Normality of standard solution (0.1 N H_2SO_4)

V = Volume of sample taken for titration (10 mL)

$$\text{Bicarbonate content (meq/L)} = (z - y) \times N \times \frac{1000}{\text{mL of sample taken (mL)}}$$

Where;

z = Titre value of the sample for methyl orange indicator

N = Normality of standard solution (0.1 N H_2SO_4)

V = Volume of sample taken for titration (10 mL)

Conclusion: The carbonate and bicarbonate of the water sample is estimated and expressed in mEq/L.

Observation Table

Sl No.	Aliquots taken (mL)	Volume of 0.1 N H_2SO_4 consumed during titration (mL)					
		Phenolphthalein			Methyl orange		
		IBR (mL)	FBR (mL)	Difference (mL)	IBR (mL)	FBR (mL)	Difference (mL)
1	10						
2	10						
3	10						

* Final burette reading of phenolphthalein becomes the initial burette reading for methyl orange

4.14 Calcium (Ca) and Magnesium (Mg) in Water Sample

Theory

The hardness of water/effluent is due to presence of mostly Ca and Mg. The extent of hardness reduces the household use of water and in irrigation water it affects the soil ecosystem. Therefore it is required to estimate the quantity of Ca and Mg present in effluent sample. Both Ca and Mg are divalent cations and are determined by titrating against EDTA (Ethylene Diamine Tetra-acetic Acid) solution, known as chelatometry or complexometric titration. A metallic indicator Eriochrome Black T (EBT) is used in this estimation.

Principle

To the water/effluent samples containing metallic cations, if the indicator is added either in the form of powder or as a solution, it combined with the cations forming a cation – indicator complex. It imparts certain colour to the solution. When the solution is titrated against strong standard chelating agent like EDTA solution, the cation indicator complexes are dissociated forming stable complex of cation with EDTA. It releases the indicator free in the solution that expresses a characteristic colour which is taken as the end point.

Magnesium forms insoluble $Mg(OH)_2$ precipitate at pH 12.0 and allows only calcium to be titrated and to increase the pH to 12.0 NaOH buffer is used. Both calcium and magnesium can be titrated at a pH 10.0 using Eriochrome Black T-as indicator. To increase the pH to 10.0 ammonia buffer is used.

Apparatus and Glassware required

Burette stand, Burette (50 mL), Pipette 5 mL, China dish (125 mL) and glass rod.

Reagents Required

1. *EDTA 0.01 N*: Dissolve 2 g of EDT A and 0.05 g of $MgCl_2$. $6H_2O$ (AR) in volumetric flask of one litre capacity and make up the volume with distilled water. Standardize the solution against 0.01 N $CaCl_2$ solution using each of the indicators, EBT and ammonium purpurate separately.
2. *Standard $CaCl_2$ 0.01 N*: Dissolve 5 g of pure $CaCO_3$ (calcite) in 10 ml of approximately 3N (1+3) HCl and add 100 mL of distilled water and boil to expel CO_2. The boiling should be done by covering the beaker with a watch glass. Rinse the watch glass and dilute to a volume of exactly one litre. This solution remains stable for many days.
3. *Ammonium purpurate (Murexide) indicator*: Grind very thoroughly 0.5 g of ammonium purpurate with 100 g of powdered K_2SO_4 (AR) in an agate motor and pestle till it becomes fine powder. This should be stored in a ground glass stoppered bottle.
4. *Eriochrome black T (EBT) indicator*: Dissolve 0.5 g EBT and 4.5 g of Hydroxylamine hydrochloride in 100 mL of 95% methanol. Store in a ground glass stoppered bottle.
5. *NaOH approximately 4 N or 16%*: Dissolve 160 g NaOH (AR) in one litre of solution and store in ground glass bottle.
6. *Ammonium chloride-ammonium hydroxide buffer*: Dissolve 67.5 g NH_4Cl (AR) in 570 mL of NH_4OH (AR) and make up to one litre.

Procedure

I. Estimation of Calcium

1. Pipette 10 mL of water sample into a dry clean 125 mL China dish.
2. Dilute with distilled water to about 25 ml.
3. Add 1 ml (15 drops) of 4N NaOH solution.
4. Add approximately 50 mg or ammonium purpurate powder. Stir the contents. The content will become vine red in colour.
5. Titrate the contents against standard EDTA (0.01 N) solution until the colour changes to violet. Since the colour change is not instantaneous, the EDTA at the end point should be added slowly at the rate of one drop for every 5 to 10 seconds. At the end point, addition of a drop of EDTA should not change the colour. To judge the end point accurately, keep a blank titrated for comparison.

6. If the sample is over titrated with EDTA, it may be back titrated with 0.01 N $CaCl_2$ till the desired end point obtained.
7. Note the volume of EDTA and repeat the titrations till consecutive and concurrent titre values are obtained.

II. Estimation of Calcium plus Magnesium

1. Pipette 10 mL of water sample in a dry clean 125 m China dish.
2. Dilute approximately to 25 mL with distilled water.
3. Add 0.5 mL (10 drops) of NH_4Cl-NH_4 OH solution.
4. Add 3 to 4 drops of EBT indicator. The colour changes to vine red.
5. Titrate against standard EDTA (0.01 N), till the colour changes to sky blue. Note the titre value (B mL).

Calculations

$$\text{Ca content (meq/L)} = T \times N \times \frac{1000}{\text{mL of sample taken (mL)}}$$

$$\text{Ca+Mg content (meq/L)} = T \times N \times \frac{1000}{\text{mL of sample taken (mL)}}$$

Mg content (meq/L) = (Ca + Mg) content - Ca content

Where:

T = Titre value i.e., volume of 0.01N EDTA solution consumed during titration (mL)

N = Normality of standard solution (0.01N EDTA) used

V = Volume of sample taken for titration (mL)

Conclusion: The calcium and magnesium content is determined and expressed in meq/L.

Observation Table for Ca or/and Ca+Mg

Sl No.	Aliquotstaken (mL)	Volume of 0.01N EDTA solution consumed (mL)			Indicator used & colour changed
		IBR(mL)	FBR(mL)	Difference (mL)	
1	10				
2	10				
3	10				

4.15 Residual Sodium Carbonate (RSC)

In irrigation waters containing high concentrations of bicarbonates (HCO_3^-) and carbonates (CO_3^{2-}) ions, there is a tendency for calcium and to a lesser extent magnesium to precipitate in the form of carbonates as the soil solution becomes more concentrated, thus leading to increase in SAR of soil solution and consequently increase in ESP of the soil. Eaton (1950) assumed that all calcium and magnesium would precipitate as carbonates and proposed the concept of "Residual Sodium Carbonates" (RSC) for evaluating high carbonate water and this can be calculated from the concentrations of carbonate, bicarbonate, calcium and magnesium.

$$RSC = (CO_3^{2-} + HCO_3^-) - (Ca^{2+} + Mg^{2+})$$

Where, the concentration of ions is expressed in meq L^{-1}.

Eaton proposed following irrigation water evaluation based on RSC values.

Table 4.4: Irrigation water evaluation based on RSC values

RSC value (me L^{-1})	*Irrigation water evaluation*
< 1.25	Safe for irrigation purpose
1.25-2.5	Marginal suitability for irrigation purpose
> 2.5	Not suitable for irrigation purpose

4.16 Chloride Toxicity and its Determination in Water Sample

Chloride Toxicity arising out of irrigation with Cl^- rich water is also common as Cl^- is not adsorbed to soils. It moves readily with soil water and is taken up by the crop, moves in the transpiration stream and accumulates in leaves. When Cl^- concentration exceeds the tolerance limit, it causes leaf burn and drying of leaf tissue. Excessive leaf necrosis is quite accompanied by early leaf drops or defoliation. In a study, on treated waste water use in agriculture with 171.44 to 254.92 mg Cl^- L^{-1} representing slight to moderate degree of restriction to the use of this waste water in irrigation.

Principle

Chloride ions present in irrigation water sample can be determined by titrating standard $AgNO_3$ using potassium chromate, as indicator, the silver of $AgNO_3$ first reacts with Cl^- in the irrigation water sample forming AgCl precipitate. Later, Ag^+ starts reacting with chromate ion (CrO_4^{-2}) to form a Reddish brown precipitate of Ag_2CrO_4. Among AgCl and Ag_2CrO_4, the solubility of AgCl is low hence it precipitates out first later Ag_2CrO_4. The very first drop of $AgNO_3$ which forms Ag_2CrO_4 is the end point of the titration.

$AgNO_3 + Cl^- \rightarrow AgCl \downarrow + NO_3^-$
(White)

$AgNO_3 + K_2CrO_4 \rightarrow Ag_2CrO_4 \downarrow + 2\ KNO_3$
(Reddish brown)

Apparatus required

Burette, Pipette, Porcelain basin and glass rod

Reagents required

1. *$AgNO_3$ (0.01 N)*: Dissolve 1.7 g of $AgNO_3$ is distilled water and make the volume to 1 li*t*.
2. *K_2CrO_4 (5 %)*: Dissolve 5 g of K_2CrO_4 in 100 ml solution.
3. 0.1 N Na_2CO_3
4. 0.1 N H_2SO_4
5. Phenolphthalein indicator

Procedure

1. Pipette out 10 ml of irrigation water sample into a porcelain basin.
2. To the contents, add 1 or 2 drops of K_2CrO_4 indicator and stir with a glass rod.
3. Titrate the contents against standard $AgNO_3$ mounted in the burette till reddish brown tinge appears.
4. Record the titre value and repeat the titrations till consecutive and concurrent titre values are obtained.

Or,

Take 50 ml of water sample in porcelain dish. Add 2 drops of phenolphthalein indicator. If pink colour appears add 0.1 N H_2SO_4 acid until solution is colourless in absence of pink colour, add 0.1 N Na_2CO_3 and then 0.1 N H_2SO_4 acid to colourless end point. Thus the pH is 8.2.Add 1 ml K_2CrO_4 indicator. Titrate against $AgNO_3$ solution until appearance of permanent reddish brown colour. A blank titration may be run simultaneously.

Calculation

$$Cl^- \text{ content (meq/L)} = T \times N \times \frac{1000}{\text{V mL of sample taken (mL)}}$$

Where;

T = Titre value i.e.; volume of 0.01 N $AgNO_3$ solution consumed during titration (mL)

N = Normality of standard solution (0.01 N)

V = Volume of sample taken for titration (mL)

Conclusion: The chloride content is determined and expressed in meq/L.

Observation Table

Sl No.	Aliquotstaken (mL)	Volume of 0.01 N $AgNO_3$ solution consumed during titration (mL)			Indicator used & colour changed
		IBR(mL)	FBR(mL)	Difference (mL)	
1	10				
2	10				
3	10				

4.17 Nitrate (NO_3) Content in Water Sample

Principle

Nitrate content in effluent sample depends on nitrogenous material present in it. NO_3 content in effluent is estimated colorimetrically. This method is based on the nitration of phenoldisulphonic acid [$C_6H_3OH(HSO_3)_2$] in fuming sulphuric acid. The colorless product $C_6H_2OH(HSO_3)_2NO_2$ (Nitrophenol disulphonic acid-in acid) behaves like nitrophenolic type indicator. It turns yellow when neutralized or made alkaline with NH_4OH due to the formation of ammonium nitrophenoldisulphonic acid ($C_6HOH(HSO_3)_2NO_2NH_2$). The intensity of the yellow color developed is estimated spectrophotometrically at 420 nm.

$$C_6H_3OH(HSO_3)_2 + HNO_3 \rightarrow C_6H_2OH\ (HSO_3)_2NO_2 + H_2O$$

$$C_6H_2OH\ (HSO_3)_2\ NO_2 + NH_4\ OH \rightarrow C_6HOH\ (HSO_3)_2NO_2NH_2 + H_2O$$

Reagent Required

1. *Nitrate stock solution 100 ppm*: dissolve 0.7215 g of AR grade KNO_3 small quantity of water and make up the volume to one litre.
2. *Working standard 10 ppm solution*: This is prepared diluting 10 ml of working standard to 100 ml.
3. *Fuming sulphuric acid*: Take 250 ml nitrate free concentration H_2SO_4 in 500 ml beaker. Keep over boiling water bath for six hours to get fuming sulphuric acid.

4. *Phenol disulphuric acid*: Take 25 g of phenol in a 500 ml beaker and add 150 ml concentrated H_2SO_4 following 75 mL fuming sulphuric acid. Keep over boiling water bath for two hours covering with a watch glass. Cool and store in an amber colored bottle.
5. *Ammonia solution*: Equal quantity of ammonia solution and distilled water are taken in a volumetric flask.

Preparation of Standard Curve

Take 0, 2, 5, 10, 15, 20 and 25 ml from 10 ppm N solution in porcelain dish. Evaporate to dryness in water bath. Then cool it and add 3 ml of phenol disulphonic acid. Add 1:1 ammonia solution slowly mixing it till the solution is alkaline as indicated by the development of yellow colour due to presence of nitrate. Add 2 ml of 1:1 ammonia solution in excess. The volume is made to 100 mL. Read the intensity of yellow colour in a spectrophotometer at 420 nm.

Procedure

1. Take 20 ml of water/effluent in a porcelain dish.
2. Evaporate it to dryness over water bath.
3. On cooling add 3 ml, phenol disulphonic acid and 1:1 ammonia solution slowly until yellow colour is observed.
4. Add 2 ml excess ammonia solution and transfer the contents to 50 ml volumetric flask. Make up the volume.
5. Intensity of yellow colour is measured in spectrophotometer at 420 nm.

Calculation

mg/litre of NO^-_3 = 2.5 x ppm read from standard curve

Where;

2.5 is the dilution factor

Conclusion: The nitrate content is determined and expressed in mg/L.

Observation Table

Sl No.	Source of water	Volume of aliquot taken (mL)	Dilution factor	Absorbance reading (A)	Concentration of (ppm)
1					
2					
3					

4.18 Sulphate (SO_4) Content in Water Sample

Principle

Sulphate (SO_4) content in water/effluent sample depends on the source from where it is collected. Usually saline effluent contains more of sulphur. It is estimated by turbidometric method. In this method effluent is treated with $BaCl_2$ powder in presence of gum acacia solution. The turbidity developed due to the precipitation of SO_4 as $BaSO_4$ is measured by a spectrophotometer at 440 nm. Gum acacia in this determination helps to prevent too rapid settling of $BaSO_4$ precipitates.

$$CaSO_4 + BaCl_2 \rightarrow BaSO_4 \downarrow (White\ ppt) + CaCl_2$$

$$MgSO_4 + BaCl_2 \rightarrow BaSO_4 \downarrow (White\ ppt) + MgCl_2$$

$$Na_2SO_4 + BaCl_2 \rightarrow BaSO_4 \downarrow (White\ ppt) + NaCl$$

Reagents required

1. *Potassium sulphate standard solution*: Weight 0.278 g AR K_2SO_4 into one litre volumetric flask. Dissolve with little water and make up the volume to one litre. This solution contains 0.05 g/ml sulphur (50 ppm S)
2. Nitric acid (6.25 %)
3. *Gum acacia* – 0.5 % gum acacia dissolved in 8.5 N acitic acid.
4. Barium chloride powder
5. *Seed Solution*: 10 ppm S in 6.25 % nitric acid

Standard curve

Pipette 0, 1, 2, 3, 4, 8, 10 ml of 50 ppm K_2SO_4 solution into 25 ml capacity volumetric flasks. Add 10 cc of 6.25 % nitric acid following 5 cc of gum acacia solution made in 8.5 N acetic acid. Add 300 g of $BaCl_2$ powder and shake for one minute. Add calculated amount of water to each flask for making volume of 25 ml. Read the standard in the spectrophotometer at 440 nm. These solutions correspond to 0, 2, 4, 8, 16, 20 ppm of sulphur.

Procedure

1. Pipette 10 ml of water sample into 50 ml test tube.
2. Add 10 cc of 6.25% HNO_3 containing 10 ppm sulphur, as seed solution.
3. Add 5 cc of 8.5 N acetic acid with 0.5 % gum acacia.
4. Add 300 mg barium chloride powder and shake for one minute to develop turbidity.

5. Carry out a blank taking distilled water as check solution and adding other chemicals as above.
6. In case of saline effluent seed solution (10 ppm solution in presence of 6.25 % HNO_3) need not to be added.
7. Measure the turbidity in spectrophotometer at 440 nm and calculate the sulphur content of the water sample from the standard curve.

Calculation

S content (ppm) = dilution factor x ppm of S read out from standard curve

$$\text{S content (mEq/L)} = \frac{\text{S content (ppm)}}{16}$$

$$SO_4 \text{ content (mEq/L)} = \frac{\text{S content (ppm)} \times 3}{16}$$

Conclusion: The sulphate is determined from the spectrophotometric values and expressed in mEq/L.

Observation Table

Sl No	Source of water	Volume of aliquot taken (mL)	Dilution factor	Absorbance reading (A)	Content (ppm)	Content (mEq/L)	Content (mEq/L)
1							
2							
3							

4.19 Sodicity Problem

Among the soluble constituents of irrigation water, sodium (Na) is considered to be most hazardous. Water which might be suitable under salinity classification may not be suitable, if sodium predominates. The effect of sodium is twofold. It affects the permeability of soil due to swelling and dispersion of clay particles and clogging the soil pores and it may cause injury to crops specifically sensitive to sodium such as fruit crops. Irrigation water containing large amount of sodium is of great concern due to plant absorbed sodium which is transported to leaves where it can accumulate and cause injury (Begum and Rasul, 2009). The sodium adsorption ratio (SAR) and residual sodium carbonate (RSC) are useful measures of sodium hazards of irrigation water (Richards, 1954).

4.19.1 Sodium Adsorption Ratio (SAR)

It is a measure of the tendency of the irrigation water to cause the replacement of calcium (Ca^{2+}) ions attached to the soil clay minerals with sodium ions which in poor soil structure and develop permeability problems. It is calculated as follows;

$$SAR_{iw} = \frac{(Na^{+})}{\sqrt{\frac{Ca^{2+} + Mg^{2+}}{2}}} \tag{4}$$

where, the ionic concentrations are expressed in miliequivalents per litre (meL^{-1}).

Table 4.5 Water quality rating based on sodium adsorption ratio (SAR)

Sodicity Problem	SAR
Low	< 10
Medium	10-18
High	18-26
Very High	> 26

Based on SAR values of irrigation water, its suitability can be evaluated with the help of rating given in the above table. In practice many irrigation water contains sufficient quantities of sulphate and bicarbonate ions to produce precipitation of calcium sulphate and calcium carbonate that remove calcium from solution and hence markedly increase sodium hazards. SAR under this situation may not give the correct sodium hazard since it is the SAR of soil water (SARsw) value rather than the SAR of irrigation water (SARiw) value that affects the ESP of the soil and an eventually the soil permeability. Hence a new index called adjusted SAR (SAR_{adj}) is suggested by Rhoades (1969 a) and is calculated by using the following formula (Ayers and Westcot, 1976).

$$SAR_{adj} = SAR_{tw}\ (1+ (8.4 + pHc)$$

The term (8.4+ *pHc*) reflects the tendency of the applied water to precipitate or dissolve $CaCO_3$. The term is defined as the theoretical pH of irrigation water with a given calcium, magnesium HCO^-_3 + CO^-_3 and + concentration, which is in equilibrium with solid $CaCO_3$. When (8.4 + pHc) > 0 calcium carbonate precipitates in the soil when the water is applied and when (8.4 + pHc) < 0, the irrigation water dissolves calcium carbonate if present in the soil. The term is calculated from the equation;

$$pHc = \left(pk_2' + pk_c'\right) + p\left(HCO_3^- + CO_3^{2-}\right)/2 + p\left(Ca^{2+} + Mg^{2+}\right)$$

Where,

$(pk'_2 + pk'_c)$ is obtained from the sum of the concentration of $Ca^{2+} + Mg^{2+} + Na^+$ in meq L^{-1} as given in Table 5. pk'_2 and pk'_c are the negative logarithms of the second dissociation constant of H_2CO_3 and the solubility-product constant of $CaCO_3$, respectively.

$P(Ca^{2+} + Mg^{2+})/2$ = negative log of concentration of calcium and magnesium in eq L^{-1}

$p(HCO_3^- + CO_3^{2-})$ = negative log of concentration in eq L^{-1}

Table 4.6: Values of for respective salt concentration in irrigation water

Sum of salt concentration (meq L^{-1})	$pk'_2 + pk'_c$
0.05-0.49	2.0
0.50-1.50	2.1
1.51-6.0	2.2
5.10-20.0	2.3
20.10-30.0	2.4
30.10-80.0	2.5

Table 4.7: Irrigation water and its suitability based on values

	Classification	Rating with suitability
< 10	S_1	Low: Suitable to black and alluvial soils, clay > 30% and fair to moderate drainage
10-20	S_2	Medium : Suitable to soils with clay 20-30 % and good drainage
20-30	S_3	High : suitable to soils with clay 10-20 % and good drainage
30-40	S_4	Very high : suitable to only light textured soils, clay < 10% and excellent drainage
> 40	S_5	Unsuitable

The value obtained by that method should be further adjusted by a 0.5 factor to evaluate more correctly the effects of HCO_3 on calcium precipitation (adj. SAR × 0.5). Soon after irrigation, dissolution or precipitation may occur, changing the supply of calcium and establishing equilibrium at a new calcium concentration different to that in the applied water. Since the SAR_{adj} does not account for these changes a new term for this is adjusted R_{Na} (Adj R_{Na}) was suggested and calculated from the following equation;

$$Adj\ R_{Na} = \frac{Na}{\sqrt{\frac{Ca_x + Mg}{2}}}$$

Where,

Na^+ = sodium in irrigation water, me L^{-1}

Ca_x^{2+} = a modified calcium concentration value taken form Table 6 in me L^{-1}.

Ca_x^{2+} represents Ca^{2+} in the applied irrigation water but modified due to salinity of the applied water, its HCO_3^-/Ca^{2+} ratio (HCO_3^- and Ca^{2+} in me L^{-1} and the estimated partial pressure of CO_2 in the surface few millimeter of soil (pCO_2 = 0.007 atm).

Mg^{2+}= Magnesium in the irrigation water me L^{-1}

The obtained is used in place of the SAR or SAR_{adj} to more efficiently evaluate the potential of water to cause an infiltration problem if used for irrigation. In a study Hameed *et al* (2010) compared SAR, SAR_{adj} and Adj R_{Na} of treated waste water and reported that SAR ranged from 1.43 to 3.19 (mean = 2.11), while and values ranged from 2.35 to 4.40 (mean = 3.12) and from 1.52 to 3.03 (mean = 2.03) respectively.

4.20 Determination of the Sodium (Na) and Potassium (K) Content in Water Sample

Principle

Concentration of sodium is quite high in saline water with EC greater than one dS/meter and relatively having less content of Ca^{2+} and Mg^{2+}. Concentration of potassium in effluent depends on the source from where it is collected. Na^+ and K^+ analysis of water/effluent sample are taken up having EC above one dS/meter at 25°C Sodium and potassium of water/effluent sample are determined with the help of Flame photometer, which is based on atomic emission spectroscopy.

When a solution of a salt is sprayed into a flame the salt breaks into the component atoms due to high temperature. The energy provided by the flame excites the orbital electrons to higher energy levels. When these electrons return to their ground state they emit a characteristic radiation. Flame emission spectroscopy are designed to measure the intensity of the characteristic line (resonance line) emitted from the excited atoms or ions (Na^+ - 589 nm, K^+ - 766 nm, Ca^{2+} - 422 nm, Mg^{2+} - 285 nm) of the element to be determined. The increase in the intensity of radiation emitted from a steady and non luminous flame will be proportional to the number of atoms of the particular element in the flame i.e., its concentration. This concentration is directly related to the content of the element in the test solution. Qualitatively Na and K can be detected from the flame colour (golden yellow or orange and pale violet respectively).

Equipments/ Glassware required

1. Flame photometer
2. Glasswares

Reagents Required

1. *1000 ppm Na stock solution*- 2.720g of NaCl (AR) dissolved in one litre of distilled water to get 1000 ppm Na.
2. *100 ppm Na working solution*- Take 10 ml of this 1000 ppm sodium solution and dilute it to 100 ml to get 100 ppm solution.
3. *1000 ppm K stock solution* - Dissolve 1.907g of KCl salt (AR) in one litre of distilled water to gives 1000 ppm K stock solution.
4. *100 ppm K working solution* - 100 ppm K^+ solution is prepared taking 10 ml of 1000 ppm K^+ solution and diluted to 100 ml.

Preparation of standard curve for Na and K

Sodium: For preparation of 0, 2, 4, 8, 16, 20 ppm of Na, take 0, 2, 4, 8, 16, 20 ml of NaCl solution from 100 ppm Na working solution and make to volume 100 ml.

Potassium: Take 0, 5, 10, 15, 20 and 25 ml of KCl solutions from 100 ppm working solution and make to volume to 100 ml to get 0, 5, 10, 15, 20 and 25 ppm of K^+.

Procedure for estimation of Na and K in effluent sample

Sodium

1. First prepare a standard curve taking the standard working solutions as described above.
2. Adjust 20 ppm solution to 100 digit of the flame-photometer.
3. Introduce the effluent sample to the flame photometer through the sucking PVC tube and observe the reading.
4. Calculate the Na content (ppm) from the standard curve prepared.

Potassium

1. Prepare the standard curve similar to sodium.
2. Adjust 25 ppm K to 100 digit of the flame-photometer.

3. Introduce the effluent sample to the flame photometer through the sucking PVC tube and observe the reading.
4. Calculate the K content (ppm) from the standard curve preapred.

Calculation

$$\text{Milliequivalent of Na}^+/\text{litre} = \frac{\text{ppm of Na}^+}{23}$$

$$\text{Milliequivalent of K}^+/\text{litre} = \frac{\text{ppm of K}^+}{39}$$

4.21 Boron (B) Toxicity

Boron, although an essential element, becomes toxic if present in excess even at relatively very low concentration of 0.6 mg L^{-1}. When boron is absorbed by the crops it tends to accumulate in the leaves until it becomes toxic to the leaf tissue and results in the death of the plant. In arid regions, boron is considered the most harmful element in irrigation water. Boron is also present in irrigation water as unionized boric acid expressed as boron element (B) in mg L^{-1}. Sensitivity to boron encompasses a wide variety of field and tree crops, although fruit, nut and berry crops are particularly very sensitive. On the basis of B content (mg L^{-1}) and the toxicity effect there are 5 classes of irrigation water (Bigger and Nielsen, 1972) as given in table 4.8.

Table 4.8: Classification of irrigation water based on boron (B) content

Boron (mgL^{-1})	Toxicity Hazard
< 0.5	Safe for sensitive crop
0.5– 1.0	Sensitive crops will show slight to moderate injury
1.0 - 2.0	Semi tolerant crops will show slight to moderate injury
2.0-4.0	Tolerant crops will show slight to moderate injury
> 4.0	Hazardous for nearly all crops

4.21.1 Determination of the Boron (B) Content in Water Sample

Boron, although an essential component for plant growth it is toxic, if present in concentration above the required limit depending upon the tolerance of the crop. Certain water/effleunt contains toxic concentration of this element and cause hazard to the crop. So prior to irrigating a crop from different sources or contamination of soil from irrigation water, boron estimation is essential. Boron estimation should be made in polythene or soft glasswares.

Reagents Required

1. *Buffer solution*: Dissolve 250 g of ammonium acetate and 15 g of disodium salt of EDTA (ethylene diamine tetra-acetic acid disodium salt) in 400 ml of deionized water and slowly add 125 ml of glacial acetic acid and mix.

2. *Azomethine-H solution*-Dissolve 0.45 g of Azomethine H in 100 ml of 1 % L-ascorbic acid solution. Every week a fresh solution should be prepared and kept in a refrigerator.

3. *Stock standard solution of B (100 ppm B)*: Dissolve 0.5716 g of boric acid (H3BO3) in a small volume of deionized water and make the final volume to 1000 ml to obtain a solution containing 100 ppm B.

4. *Working standard solution of B (20 pm)*: Dilute the stock solution by 5 times i.e., take 20 ml of 100 pm and make it to 100 ml with deionized water and mix.

Equipment required

1. Spectrophotometer
2. Glasswares

Preparation of standard curve

1. Pipet out 0, 1, 2, 3, 4 and 5 ml of the working standard solution of B into 100 ml volumetric flasks and make the volume with deionized water to obtain final standard solutions containing 0, 0.2, 0.4, 0.6, 0.8 and 1 ppm B respectively.

2. Transfer 1 ml each of the final standard solutions into a separate polyethylene tubes or a small beakers. Add 2 ml buffer solution and mix the content thoroughly. Then add 2 ml azomethine-H reagent and then mix the contents thoroughly. Allow the mixture to stand for 30 minutes and measure the intensity of yellow colour developed using a spectrophotomer at 420 nm.

Procedure for estimation of boron

1. Take one ml of water/effluent sample into a polyethene test tube or 50 ml polythene beaker.

2. Add 2 ml of buffer solution following 2 ml azomethine- H and mix the contents thoroughly.

3. Measure the intensity of a yellow colour developed after 30 minutes through a spectrophotometer at 420 nm.

4. Record the absorbance of the coloured solution.

Calculation

Boron content (ppm)

= dilution factor x concentration of B read out from the standard curve in ppm

Table 4.9: Classification of crops according to Boron

Boron class	Sensitive crops	Semi tolerant crops	Tolerant crops
B_1	Less than 0.33	Less than 0.67	Less than 1.00
B_2	0.33-0.67	0.67-1.33	1.00-2.00
B_3	0.67-1.00	1.33-2.00	2.00-3.00
B_4	1.00-1.25	2.00-2.50	3.00-3.75
B_5	Above 1.25	Above 2.50	Above 3.75

Effluent falling under B_1 and B_2 classes can be safely used for irrigation. While those under B_3 and B_4 would be of marginal quality and B_5 would be toxic to crop.

Table 4.10: Relative tolerance of crops to boron

Sensitive	Semi-tolerant	Tolerant
Plum	Potato	Palm
Pear	Cotton	Date palm
Apple	Tomato	Sugar beat
Grape	Peas	Onion
Cherry	Radish	Turnip
Peach	Barley	Cabbage
Orange	Wheat	Lattuce
Grape fruit	Oat	Carrot
Lemon	Pumpkin	Cauliflower

4.22 Other Trace Elements

In some areas, toxic levels of selenium occur in well water. It has been reported that lithium in well water is toxic to citrus. Industrial pollution of natural water with copper, nickel, cadmium and other toxic heavy metals can also occur. These inputs to the environment are manmade and can be controlled by cleaner technology and by strict guidelines. There are two important guidelines for irrigation water quality; 1. Even if toxicity levels of the element remain below accepted irrigation water quality standard, it accumulates in the soil of irrigated crop field if output is below input, 2. Excessive concentration in soil and water has adverse impacts on crop growth and development that leads to immediate hazard to the aquatic and terrestrial lives. Hence sustainability principles are required to minimize this accumulation of toxic elements in agricultural soils.

The accumulation of a given substance ($M_{accumulation}$) in irrigated land can be expressed;

$$M_{accumulation} = \sum_{i=1}^{n} \Delta M_i$$

Where,

ΔM_i = the change in concentration of the substance in the soil in a specific time interval (i), which can be explained by the simplified mass-balance equation;

$$\Delta M_i + M_{in} - M_{drain} - M_{grain} - M_{process}$$

Where,

M_{in} = the mass of the substance imported to the field through irrigation water, fertilizer, atmosphere or other sources.

M_{drain} = the mass exported from the field through surface and subsurface drainage.

M_{grain} = the mass taken up by crops and via the grain exported out of the system.

$M_{process}$ = relevant in the case of e.g. pesticides where a substance can disintegrate into other components due to temperature, sunlight exposure, bacterial activity etc.

It is observed from the above equation that ÄMi must be equal to/or less than zero to avoid accumulation in the soil. It also indicates that to lower $ÄM_i$ either a reduction in M_{in} or an increase in M_{drain} or M_{grain} has to be achieved. M_{drain} or M_{grain} can to a certain degree be controlled either by improved drainage or by specific crop selection. $M_{process}$ is on the other hand more dependent on the climate and soil conditions and therefore difficult to control. Guidelines concerning the maximum permissible levels of various trace elements in irrigation water have been reported in table 10 (FAO, 1985). Siddiqui (1995) also reported the same value for Zn, Mn, Pb, Ni, Co, and Cd to judge the suitability of irrigation water. Besides he reported 0.2 mgL^{-1} for copper (Cu) and 0.5 mgL^{-1} for iron (Fe) as the maximum limit for irrigation.

4.22.1 Estimation of Heavy/ Trace Metals in Water Samples

Heavy metals are generally defined as metals with relatively high densities, atomic weights or atomic numbers. Density criteria range from above 3.5 g/cm^3 to above 7 g/cm^3. Chromium, arsenic, cadmium, mercury and lead have the greatest potential to cause harm on account of their extensive use. In the aquatic environment a heavy metal can occur in several forms: as a free ion –

the form most toxic to living organisms; as an ion bound to different ligands – complex compounds are formed; as the precipitated molecule of a compound, suspended in the liquid phase or adsorbed on the surface of suspended or colloidal matter.

For the determination of trace metals, contamination and loss are of prime concern. Most metals present in water in the dissolved state, as well as those in the other two forms, after suitable treatment are brought into the dissolved state, can conveniently be determined by means of atomic absorption spectrophotometry. In an atomic absorption analysis, the element being determined must be reduced to the elemental state, vaporized and imposed in the beam of radiation from the light source. This process is most frequently accomplished by drawing a solution of the sample, as a fine mist, into a suitable flame. The flame thus serves a function analagous to that of the cell and solution in conventional absorption spectrophotometry. The absorption is measured at a selected wavelength which is characteristic for each individual element. The absorbance measured is proportional to the concentration and analysis is carried out by comparing this absorbance with that given under the same conditions by reference samples of known composition.

Apparatus Required

Atomic Absorption Spectrophotometer,

Glass wares and Filter paper

Reagents Required

1. Primary Standard solutions of Cd, Cr, Pb, Ni, Fe, Mn, Zn, Cu, As, Hg, Co : 1000 ppm stock solution from pure analytical grade
2. Secondary standards : 100 ppm from the primary standards
3. Working standards: As per specifications of Instruments and it should be prepared a fresh.
4. Conc. HNO_3

Procedure

1. The sample container, whether borosilicate glass, polyethylene, polypropylene or teflon, should be thoroughly washed.
2. Samples should be preserved immediately after sampling by acidifying with concentrated nitric acid to pH < 2. (Addition of 10 mL of concentrated nitric acid per litre of sample will normally be sufficient).

3. After acidification, the sample should be stored in a refrigerator at approximately 4 °C.
4. Under these conditions, samples with metal concentrations of several milligrams per litre are suitable for up to 6 months (except mercury, for which the limit is 38 days in glass and 14 days in plastic).
5. During analysis the samples are filtered using Whatman No 42 filter paper.
6. Filtrate is directly fed to AAS for determination of different heavy metals or after necessary dilution and results are expressed in ppm or mg/L.

Calculation

Metal Concentration in effluent/ water = AAS reading in ppm X dilution factor (if any)

Table 4.11: Recommended maximum concentration of Trace Elements in irrigation water (FAO, 1985)

Elements	For water used continuously on all soils (mgL^{-1})
Aluminium	5.0
Arsenic	0.1
Beryllium	0.1
Cadmium	0.01
Chromium	0.1
Cobalt	0.05
Fluorine	1.0
Iron	5.0
Lead	5.0
Lithium	2.5
Manganese	0.2
Molybdenum	0.01
Nickel	0.2
Selenium	0.02
Zinc	2.0

High concentration of heavy metals in irrigation water can result in death of crops, interface with uptake of other essential nutrients or form objectionable deposits in fruits and render edible portion of plants toxic to human and grazing animals.

4.23 Microbiological Analysis of Water Sample

The microbiological examination of water emphasizes assessment of the hygienic quality of water supply. It requires the isolation and enumeration of organisms that indicate the presence of contamination. For selecting water samples for microbiological studies, the following criteria are usually considered;

Table 4.12: FAO guidelines for evaluation of irrigation water quality

Sl No.	FAO guidelines for evaluation of irrigation water quality				
	Soil property affected	Units	Water quality guidelines		
			No problem	Increasing problem	Severe problem
1	Crop water availability	EC_{iw} (dS/m)	0.7	0.7 – 3.0	> 3.0
2	Permeability (Adj. SAR)	Montmorillonite	< 6.0	6.0 – 9.0	> 9.0
		Illite – Vermiculite	< 8.0	8.0- 16.0	> 16
		Kaolinite – Sexquioxide	< 16.0	16.0- 24.0	>24
3	Specific ion toxicity for sensitive crops	Sodium (Na) –Adj. SAR	< 3.0	3.0 – 9.0	>9.0
		Chloride (Cl) – me/L	< 4.0	4.0 – 10.0	> 10.0
		Boron (B)- mg/L	< 0.7	0.7 – 2.0	> 2.0
4	Miscellaneous effect for susceptible crops with sprinkler and drip	[illegible] $_3^-$ or NH_4^+) – mg/L	< 5.0	5.0 – 30.0	> 30.0
		Bicarbonate(HCO_3^-) - me/L	< 1.5	1.5 – 8.5	> 8.5
5	pH (may cause imbalance in nutrient uptake)	pH	6.5 – 8.5		Very low or very high

Source: FAO,1985

- The point of sampling should be selected in such a way that the samples taken are representative of the different sources.
- Sampling points should be uniformly distributed throughout a distribution system.
- Water should be sampled from the reservoirs.
- In systems with more than one water source, the locations of the sampling points should take account of the number of inhabitants served by each source.

4.23.1 Bacteriological analysis of water sample

The microbiological examination of water emphasizes assessment of the hygienic quality of water to know the presence of infectious disease causing organisms. This requires the isolation and enumeration of organisms that indicate the presence of faecal contamination. In certain circumstances, the same indicator organisms may also be used to assess the efficiency of drinking-water treatment plants, which is an important element of quality control. Other microbiological indicators, not necessarily associated with faecal pollution, may also be used for this purpose. The isolation of speciûc pathogens in water should be undertaken by reference laboratories.

4.23.1.1 Isolation of *Enterococous* bacteria in water sample

It is determined by membrane-ûltration (MF) method. In this method a volume of 10-100ml of the sample is introduced aseptically into a sterile ûltration assembly containing a sterile membrane ûlter (nominal pore size 0.2 or 0.45µm). A vacuum is applied and the sample is drawn through the membrane ûlter. All indicator organisms are retained on or within the ûlter, which is then transferred to a suitable selective culture medium in a Petri dish. It is then incubated for a suitable time to allow the replication of the indicator organisms. The colonies formed are counted and the results are expressed in CFU (colony forming units) per 100 ml of original sample. This technique is not suitable for water samples having high tubidity as it blocks the pores of the filter.

Requisites

1. Membrane filter apparatus
2. Sterile glasswares, needles
3. Laminar air flow
4. BOD incubator

5. *Enterococcus* Agar base media
6. Vaccum pump

Composition of *Enterococcus* Agar Base Media

Ingredients	g/L
Casein enzymic hydrolysate	15.0
Papaic digest of soyabean meal	5.0
Yeast extract	5.0
Dextrose	2.0
Dipotassium phosphate	4.0
Sodium azide	0.4
2,3,5-Triphenyl tetrazolium chloride	0.1
Agar	10.0
Final pH	7.0

Procedure

1. Take 10 mL of the water sample and filter it through the membrane-filtration (MF) aseptically.
2. A vacuum is applied and the sample is drawn through the membrane filter. The organisms retained on or within the filter are then transferred to a petridish containing *Enterococcus* Agar Base Media.
3. The petri dish with organisms is then incubated in a BOD incubator at 28-30°C for multiplication of the organisms.
4. After incubation, the colonies formed are counted and the results are expressed in CFU per 100 ml of water sample.

Calculation

$$\text{Total number of colonies developed (cfu/100 ml)} = \frac{\text{Coliform colonies formed}}{\text{Volume of sample filtered}} \times 100$$

4.23.2 Isolation of *Escherichia coli* bacteria in water sample

It is possible to remove bacteria from fluids by passing them through filters with such small pore size that bacteria are arrested. This filtration technique enables fairly large volumes of water to pass rapidly under pressure, but prevents the passage of any bacteria present. These organisms are then retained on the surface of the membrane filter which is then brought into contact with suitable liquid nutrients media for multiplication and counting.

Requisites

1. Membrane filter apparatus
2. Sterile glasswares, needles
3. Laminar air flow
4. BOD incubator
5. Endo Agar LES base media
6. Lauryl Sulphate Broth
7. Vaccum pump

Composition of M-Endo Agar LES

Ingredients	g / L
Casein enzymic hydrolysate	3.7
Peptic digest of animal tissue	3.7
Tryptose	7.5
Yeast extract	1.2
Lactose	9.4
Dipotassium phosphate	3.3
Monopotassium phosphate	1.0
Sodium chloride	3.7
Sodium deoxycholate	0.1
Sodium lauryl sulphate	0.05
Sodium sulphite	1.6
Basic fuchsin	0.8
Agar	15.0
Final pH (at 25°C)	7.0

Composition of M-Lauryl Sulphate Broth

Ingredients	g / L
Peptic digest of animal tissue	40.0
Yeast extract	6.0
Lactose	30.0
Sodium lauryl sulphate	1.0
Phenol red	0.2
Final pH (at 25°C)	7.4±0.2

Procedure

1. Take 10 mL of the water sample and filter it through the membrane-ûltration (MF) aseptically.

2. A vacuum is applied and the sample is drawn through the membrane ûlter. The organisms retained on or within the filter are then transferred to a petridish containing Endo Agar LES base media and then impregnated with 2 ml Lauryl Sulphate Broth.

3. The petri dish with organisms is then incubated in a BOD incubator at 35 ^{0}C for 1 - 1½ hours for multiplication of the organisms.

4. After incubation, the colonies formed (coliforms produce golden green colonies with metallic sheen within 24 hours of incubation) are then counted and the results are expressed in CFU per 100 ml of water sample.

Calculation

Total number of coliform colonies developed (cfu/100 ml)

$$= \frac{\text{Coliform colonies formed}}{\text{Volume of sample filtered}} \times 100$$

Suggested Readings

Baruah, T.C. and Barthakur, H.P. A text book of soil analysis, 1995. Vikas Publishing House, New Delhi.

Eaton, F.M., Significance of carbonates in irrigation waters. *Soil Science*, 1950, **69**: 123-133.

F.A.O., Water quality for agriculture, R.S. Ayers and D.W. Westcot, Irrigation and drainage paper, FAO, Rome, paper 29, Rev. 1, 1985, 174 p.

Jackson, M.L. (1973) Soil Chemical Analysis, Prentice Hall of India Pvt. Ltd., New Delhi.

Kanwar, J.S. and Chopra, S.L. (1988) Analytical Agricultural Chemistry, third Edition, Kalyani Pub., New Delhi.

Lenka, D., Irrigation and Drainage, Kalyani Publisher, India, 1991.

Oster, J.D. and Rhodes, J.R., Various indices for evaluating the effective salinity and sodicity of irrigation waters, In *Proc. Int. Salinity Conf.*, Lucbock Texass, 1977.

Oster, J.D. and Schroer, F.W., Infiltration as influenced by irrigation water quality. *Soil Sci. Soc. Am. J.*, 1979, **43**: 444-447.

Page, A.L., Miller, R. H. and Keeney, D.R. Methods of Soil Analysis, 1982, 2nd Edition, Madison Wisconsin, USA.

Paliwal, K.V., Irrigation with saline water. IARI, New Delhi, 1972, Monograph No. 2.

Pescod M.B., Wastewater Treatment and Use in Agriculture: Irrigation and Drainage paper No. 29, 1985, FAO, rome.

Rhodes, J.R., Mineral weathering correction for estimating the sodium hazard for irrigation waters. *Soil Sci. Soc. Am. Proc.*, 1968, **32**: 648-552.

Richards, L.A., Diagnosis and improvement of saline and alkali soils, USDA, Washington DC, 1954, Agric. Handbook No. 60.

Shainberg, I. and Oster, J.D., Quality of irrigation water. I.I.I.C. publication No.2, 1978, (Inst. Irrigation Inf. Centre, Israel).

Siddiqui, K.A., Pollution conservation and Forestry: for undergraduate students, 1995, Kitab Mahal, New Delhi

Singh, D. Chhonkar, P.K. and Dwivedi, B.S. Manual on Soil, Plant and Water Analysis, 2007. pp 199. Westville Publishing House, New Delhi.

Szabolcs, I and Darab ,C. The influence of irrigation water of High Sodium Carbonate content of soils, *Proceedings of 8th International Congress of ISSS*, 1964, **2**: 803-812.

WHO, Guidelines for the safe use of waste water, Excreta and Grey water in Agriculture, World Health Organization, Geneva, 2006, Vol. 2.

Appendices

A. Rating chart of soil test values of primary nutrients

Nutrients	Rating*		
	Low	Medium	High
Organic carbon (%)**	$\leq$ 0.5	0.51 to 0 .75	> 0.75
Alkaline $KMnO_4$-N (kg/ha)	$\leq$ 280	281- 560	> 560
Olsen's P (kg/ha)	$\leq$ 10	11 – 25	> 25
Amm. Acetate K (kg/ha)	$\leq$ 120	121- 280	> 280

* Subject to minor variation as per local conditions.
** As an index of available N

B. General recommended dose of micronutrient fertilizers

Micronutrients	*Materials and doses for application*	
	Soil application	*Foliar application*
Zinc	Zinc sulphate (25 kg/ha)	0.5 % zinc sulphate + 0.25 % lime
Iron	Ferrous sulphate (50 kg/ha)	1% Ferrous sulphate + 0.5 % lime
Copper	Copper sulphate (10kg/ha)	0.1 % copper sulphate + 0.05 % lime
Manganese	Manganese sulphate (10 kg/ha)	1 % manganese sulphate + 0.25 % lime
Boron	Borax (10 kg/ha)	0.2 % Borax

C. Soil test based fertilizer schedule for NPK***

N		P		K	
Org. Carbon (%)	N (kg/ha) *	Olsen's P (kg/ha)	P_2O_5 (kg/ha)*	Ammo. Acetate K (kg/ha)	K_2O (kg/ha)*
$\leq$ 0.20	120©	$\leq$ 10	80	$\leq$ 100	60
0.21 – 0.40	120®	11-20	60	101-150	50
0.41 -0.60	120	21-30	40	151-200	40
0.61 -0.80	90	31-40	30	201-250	30
0.81 -1.00	75	41-50	20	251-300	20
>1.00	60	>50	10	>300	0

*** for HYV of wheat, rice and maize under irrigated condition.
* Amount to be applied; © Along with 10 t/ha of FYM; ® Along with 20 t/ha of FYM

D. Concentration, Specific gravity and normality of analytical reagents

Analytical Reagents	Concentrations (%, w/w)	Specific gravity (g/mL) at 25 °C	Normality
Acetic acid glacial	99.0	1.06	17.5
Sulphuric acid	96.0	1.84	36.0
Hydrochloric acid	39.0	1.19	11.8
Nitric acid	71.0	1.42	15.6
Phosphoric acid	85.0	1.70	44.0
Ammonium hydroxide	28.3	0.91	15